T0259713

Notes on the Ellipsoidal Function

Chang-Hong Liang

Notes on the Ellipsoidal Function

Chang-Hong Liang
School of Electronic Engineering
Xidian University
Xi'an, Shaanxi
China

The book has been published in Chinese language by Science Press, Beijing, in Oct. 2014. ISBN 978-7-03-042160-9 Science Press has agreed to publish the English version and the cooperation details will be discussed.

ISBN 978-981-10-9737-9   ISBN 978-981-10-2908-0   (eBook)
DOI 10.1007/978-981-10-2908-0

Jointly published with Science Press, Beijing

Printed on acid-free paper

This Springer imprint is published by Springer Nature
The registered company is Springer Nature Singapore Pte Ltd.
The registered company address is: 152 Beach Road, #22-06/08 Gateway East, Singapore 189721, Singapore

Foreword

The circle, ellipse, sphere, and ellipsoid are the geometric forms we frequently encounter, and in scientific research or engineering technology we must often deal with many problems related to such forms. For example, the earth, due to its complicated shape, is replaced by the relatively simpler "reference ellipsoid" in geodesy to carry out various calculations, demarcate the coordinates of longitude and latitude, and mark elevations on its surface. A famous classical problem in electromagnetics is the calculation of the electric field and magnetic field caused by a point charge moving at a constant speed. Heaviside derived relevant equations as early as 1888, and subsequently others conducted similar calculations under different assumptions or through different approaches. Despite their different methods, they had a common conclusion that the electric field was contracted or deformed along the direction of motion into an ellipsoid, which was then called a Heaviside ellipsoid. In discussing the problems of electromagnetic wave propagation and optics, we often come across anisotropic materials. If a small block of such material is placed in a rectangular coordinate system, with its edges coinciding with the axes X, Y and Z respectively, the square roots of dielectric constant along these directions stand for electric displacement vector's refractive indexes of electromagnetic waves along the corresponding axes. An ellipsoid drawn with those refractive indexes as axial lengths is the famous Fresnel ellipsoid, or the ellipsoid of refractive indexes, which can conveniently describe the features of an electromagnetic wave when it propagates along each direction of the material. There are even more examples of elliptic or cylindroid structures. What researchers of electromagnetics first think of are definitely the fundamental works concerning the elliptic cross-sectional waveguide proposed by Lan Jen Chu in 1938, as well as various component products of elliptic waveguide that emerged in the next 10 years. It is needless to mention the significance of elliptic function in mathematical and physical approaches.

Compared with the simple and mature functions and theories of the circle and sphere, the study of ellipse and ellipsoid is much more complicated, so beginners need more guidance and help. Thus, I am delighted to see Prof. Chang-Hong

Liang's new writing, *Notes on the Ellipsoidal Function*. The book is original in the arrangement of its content for readers who are engaged in the study of electromagnetic fields. The book starts with the simplest circular and trigonometric functions, gradually shifts to elliptic integrals, and finally guides the reader to a primary knowledge of ellipsoidal integrals. After the shift to elliptic integrals, the author turns to talk about several concrete electromagnetic examples in the ellipse and ellipsoid, and makes comparisons and mutual verifications through the method of moment. The purpose is to ensure a close connection between electromagnetics and mathematics. The author then initiates another upsurge of learning mathematics, elaborating on theories of ellipsoidal function and its conformal mapping, and ultimately applies them to several practical electromagnetic problems such as the ellipsoidal function network and filter. The arrangement completely matches the thinking modes and knowledge structure of senior undergraduates and junior graduates majoring in related programs. With concise characters and clear narration, the book is an ideal reading material for engineering mathematics.

Complete, rigorous theory is necessary in order to deal with concrete problems through ellipsoidal function. However, the task also calls for extensive numerical calculations, charts and curves. The theories revealed in this book will, together with widely used practical tools such as MATLAB, help readers achieve great success.

This book is another work by Prof. Chang-Hong Liang in the engineering mathematic field, following those such as *"Notes on Vector Field Theory," "Notes on Complex Variable Function,"* etc. With great perseverance and consistency, he has demonstrated a meticulous approach in his scholarship. He has continued to write and has made the remarkable achievements in the past decades. I believe this book will again push forward the research in the field of electromagnetics, and I congratulate him on his prominent academic achievements.

Pei-Heng Wu
Fellow of Chinese Academy of Science
Professor of Nanjing University

Preface

This book is my fifth volume on engineering mathematics, following *Notes on Vector Field Theory* (Science Press, 2007), *Notes on Complex Variable Function* (Science Press, 2011), *Notes on Matrix Theory* (Science Press, 2014), and *Notes on Probability Theory* (Science Press, 2014). Although each note touches on a completely different field, *they all serve the same goal: endeavoring to build a bridge between mathematics and engineering while bringing about benefits for readers in their future learning and work.*

The core theme of the book is ellipsoid function, which serves as an extension of the existing circular function (trigonometric function) and elliptic function. If circular function, elliptic function and ellipsoidal function are compared to three peaks, each of which is higher and more magnificent than the preceding one.

Ellipsoid function belongs to the special function field. The mention of special function would discourage a good many readers due to its abstraction and complexity. The first feature of the book, or the most striking one, is the elicitation of the most complicated ellipsoid function from the most practical and specific application examples. Moreover, the book begins with the discussion from the elliptic integral.

Ellipsoid integrals fall into two categories. Here we take the first category as the example. Circular integral corresponds to the capacitance of the circular disk conductor; elliptic integral of the first kind corresponds to the capacitance of the elliptic disk conductor; while ellipsoid integral of the first kind corresponds to the capacitance of the ellipsoidal conductor. This form of discussion is accessible and concrete, with each being encompassed by another, namely:

Ellipsoid integral of the first kind
Elliptic integral of the first kind
Circular integral

Let us consider the second category of ellipsoid integrals. Circular integral corresponds to the perimeter of a circle; elliptic integral of the second kind corresponds to the perimeter of an ellipse; and ellipsoid integral of the second kind corresponds to the surface area of an ellipsoid. It is obvious that each of them encompasses another as well, namely:

Ellipsoid integral of the second kind
Elliptic integral of the second kind
Circular integral

The second feature of the book is the close combination of algebra and geometry. The book discusses the ellipsoid function (algebra) theory and conformal mapping (geometry) theory of the ellipsoid function in exhaustive detail. The latter is the complex geometry theory of ellipsoid function as opposed to the ordinary geometrical theory, which creates the most illustrative background for the application of the function.

The third feature of the book, namely the major goal of engineering mathematics, is the application to solving practical engineering problems. In the book, I set up the ellipsoid function network and three-band filter (pass band, transition band and attenuation band) model, ultimately producing an ellipsoid function filter with outstanding performance through concrete examples. In other words, we delineate the path that engineering mathematics should take, using the ellipsoid function as the example.

Additionally, I take the liberty of discussing the core idea of the serial books. Entrepreneurship, innovation, and creation have become very fashionable terms. However, it should be pointed out that they are not just a form of empty talk or an ornament. Instead, they represent the attitude, ambition, and persistence that we should hold and bear in mind. Innovation is valuable, regardless of whether it is big or small. I happened to follow the idea in writing the series of books. This is also true of, for example, vector division, complex operator, and polarization networks. Open ideas, varied contents, and popular characters are my constant pursuits in writing the books.

Despite my painstaking efforts, it is inevitable that there might be some shortcomings in the book. Therefore, I sincerely hope that experts and readers can make comments and corrections to further enhance the quality of the serial books.

Finally, I extend my heartfelt thanks to Prof. Pei-Heng Wu, a respected member of Chinese Academy of Sciences, for his help and preface to the book.

Xi'an, China   Chang-Hong Liang

Contents

About the Author

 **Chang-Hong Liang** born in Shanghai, China in December 1943, is a professor, doctoral supervisor and senior member of IEEE. He graduated from the Department of Physics of People's Liberation Army Institute of Telecommunication Engineering (the current Xidian University) in 1965, and took a teaching post there after receiving a master's degree from the college in July 1967. He was invited to serve as a visiting scholar by Syracuse University from 1980 to 1982, and was appointed as the President of Xidian University from 1992 to 2002. He has long been engaged in the study of microwaves and electromagnetism, and has made the substantial achievements, particularly in computational microwave, nonlinear electromagnetism, and microwave network theories. Moreover, he has won a dozen provincial and ministerial scientific and technological awards as well as teaching awards, and has published a dozen monographs and translations. Being meticulous in scholarship and a model of virtue for his students, he continued to teach basic courses even after he became the president of the university. His lectures were vivid but readily understandable. In 2003, he obtained the first "National-grade Famous Teacher" Prize awarded by Ministry of Education of China, and his lecture "Fundamentals of Microwave Technology" was evaluated as the first "National Superior Course" in the same year.

Chapter 1
From Circle Integral to Ellipsoidal Integral

1.1 Introduction

Circle, ellipse, and ellipsoid are three commonly encountered geometrical graphs in engineering, as shown in Fig. 1.1.

First, we observe the difference among three above graphs in geometric dimension. Obviously, circle and ellipse are two-dimensional graphs, while ellipsoid is a three-dimensional graph.

Second, from the viewpoint of the symmetry, we know that a circle is a symmetric two-dimensional graph, and an ellipse is not a symmetric one. Moreover, an ellipsoid is not a symmetric three-dimensional graph (the symmetric graph is a sphere in three-dimensional space).

The algebraic equations of three above graphs shown in Fig. 1.1 are

$$x^2 + y^2 = R^2 \text{ (circle)} \tag{1.1}$$

$$\frac{x^2}{a^2} + \frac{y^2}{b^2} = 1 \text{ (ellipse)} \tag{1.2}$$

$$\frac{x^2}{a^2} + \frac{y^2}{b^2} + \frac{z^2}{c^2} = 1 \text{ (ellipsoid)} \tag{1.3}$$

We list the characteristic geometry variables of three graphs in Tables 1.1 and 1.2. It can be clearly seen that the advanced idea about new integral of some kind is proposed in the research procedure of characteristic geometry variables of ellipse and ellipsoid, which is also a key point in this book. As an aside, in the ancient time of china, the research on π results from the characteristic geometry variables of circle.

© Science Press, Beijing and Springer Nature Singapore Pte Ltd. 2017
C.-H. Liang, *Notes on the Ellipsoidal Function*,
DOI 10.1007/978-981-10-2908-0_1

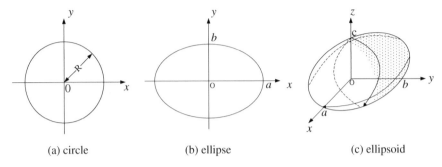

(a) circle   (b) ellipse   (c) ellipsoid

Fig. 1.1 Geometrical graphs of circle, ellipse, and ellipsoid

Table 1.1 Characteristic geometry variables of circle and ellipse

	Circle	Ellipse
Perimeter	$2\pi R$	Elliptic integral of the second kind
Area	πR^2	πab

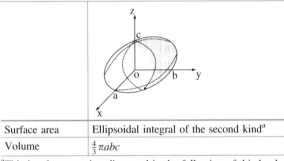

Table 1.2 Characteristic geometry variables of ellipsoid	
Surface area	Ellipsoidal integral of the second kind[a]
Volume	$\frac{4}{3}\pi abc$

[a]This is a key question discussed in the following of this book

1.2   From Circle and Elliptical Integrals to Ellipsoidal Integral

Trigonometric functions are a kind of well-known functions, which have been widely used in practical engineering applications. They are initially defined by a unit circle, as shown in Fig. 1.2.

Fig. 1.2 Definitions of
trigonometric functions in a
unit circle

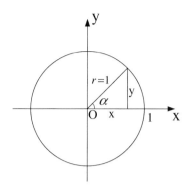

Here let us directly write the definitions of trigonometric functions in the first quadrant as follows:

$$\sin \alpha = \frac{y}{r} \tag{1.4}$$

$$\cos \alpha = \frac{x}{r} \tag{1.5}$$

$$\tan \alpha = \frac{y}{x} \tag{1.6}$$

The above definitions in the first quadrant are without loss of generality. Meanwhile, the definitions do not depend on whether r is equal to 1. In other words, it has no effect on whether the unit circle is used.

$$\sin^2 \alpha + \cos^2 \alpha = 1 \tag{1.7}$$

$$1 + \tan^2 \alpha = \frac{1}{\cos^2 \alpha} \tag{1.8}$$

It is worthwhile pointing out that trigonometric functions are called circle functions due to the circle-based definitions. On the other hand, $\sin \alpha$ is a fundamental function, and other functions can be derived according to $\sin \alpha$. With the assumption of α in the first quadrant, we can easily write

$$\cos \alpha = \sqrt{1 - \sin^2 \alpha} \tag{1.9}$$

$$\tan \alpha = \frac{\sin \alpha}{\sqrt{1 - \sin^2 \alpha}} \tag{1.10}$$

In this book, trigonometric functions, i.e., circle functions are defined in terms of integral from a new viewpoint, which is different from the above conventional definitions of trigonometric functions. As we all know

$$\int_0^x \frac{dt}{\sqrt{1-t^2}} = \sin^{-1}(x) \tag{1.11}$$

The new definitions of circle functions can be given according to Eq. (1.11).

Definition Sine circle function $\sin \alpha$ can be defined as

$$\alpha = \int_0^x \frac{dt}{\sqrt{1-t^2}} = \sin^{-1}(x) \tag{1.12}$$

Therefore, we have

$$x = \sin \alpha \tag{1.13}$$

$$\cos \alpha = \sqrt{1 - x^2} \tag{1.14}$$

$$\sin^2 \alpha + \cos^2 \alpha = 1 \tag{1.15}$$

These definitions are called circle functions. Similarly, these are easily generalized to elliptical integral.

Definition Ellipse sine function is denoted by $S_n(u, k)$.
Given a general elliptical integral

$$u = \int_0^x \frac{dx}{\sqrt{1 - x^2}\sqrt{1 - k^2 x^2}} \tag{1.16}$$

in which elliptical parameter is given by

$$0 \le k = \frac{\sqrt{a^2 - b^2}}{a} \le 1 \tag{1.17}$$

Conventionally, in Eq. (1.17) $a \ge b$ is assumed. We denote

$$x = S_n(u, k) \tag{1.18}$$

$$C_n(u, k) = \sqrt{1 - x^2} \tag{1.19}$$

Equations (1.18) and (1.19) satisfy

$$S_n^2(u,k) + C_n^2(u,k) \equiv 1 \tag{1.20}$$

Further, the ellipse Δ function $d_n(u,k)$ is defined as

$$d_n(u,k) = \sqrt{1 - k^2 x^2} \tag{1.21}$$

We have

$$k^2 S_n^2(u,k) + d_n^2(u,k) \equiv 1 \tag{1.22}$$

It is worthwhile pointing out that the above definitions are well-established Jacobi elliptical integrals. Compared with the case of $k = 0$, i.e., $a = b$, the elliptical integrals are reduced to the circle integrals as follows:

$$u(k=0) = \int_0^x \frac{dx}{\sqrt{1-x^2}} \tag{1.23}$$

$S_n(u,k)$, $C_n(u,k)$, and $d_n(u,k)$ introduced above are called Jacobi ellipse sine function, Jacobi ellipse cosine function, and Jacobi ellipse Δ function, respectively.

More precise general definitions can be summarized as follows:

Complete elliptic integral of the first kind is defined as

$$K(k) = \int_0^1 \frac{dx}{\sqrt{1-x^2}\sqrt{1-k^2 x^2}} \tag{1.24}$$

General elliptic integral of the first kind is defined as

$$u(x,k) = \int_0^x \frac{dx}{\sqrt{1-x^2}\sqrt{1-k^2 x^2}} \tag{1.25}$$

Complete elliptic integral of the second kind is defined as

$$E(k) = \int_0^1 \frac{\sqrt{1-k^2 x^2}}{\sqrt{1-x^2}} dx \tag{1.26}$$

General elliptic integral of the second kind is defined as

$$v(x,k) = \int_0^x \sqrt{\frac{1-k^2 x^2}{1-x^2}} dx \tag{1.27}$$

Based on practical requirement, we further introduce complementary elliptic integral as

$$F(k) = \int_0^1 \frac{x^2 dx}{\sqrt{1-x^2}\sqrt{1-k^2x^2}} \tag{1.28}$$

Combined with Eq. (1.26), we can obtain

$$E(k) = \int_0^1 \frac{1-k^2x^2}{\sqrt{1-x^2}\sqrt{1-k^2x^2}} dx$$

$$= \int_0^1 \frac{dx}{\sqrt{1-x^2}\sqrt{1-k^2x^2}} - k^2 \int_0^1 \frac{x^2}{\sqrt{1-x^2}\sqrt{1-k^2x^2}} dx \tag{1.29}$$

Hence we can get

$$E(k) = K(k) - k^2 F(k) \tag{1.29}$$

Two kinds of different ellipse integrals are tightly connected by the introduction of $F(k)$.

In the following, we preliminarily discuss the geometric meaning of the elliptic integrals. Figure 1.3 shows the geometric interpretation of circle integrals, e.g., trigonometric functions. Figure 1.3 clearly demonstrates the definitions of Eqs. (1.13), (1.14), and (1.15). By comparison, we give the geometric interpretation of elliptic integrals, as shown in Fig. 1.4.

According to Fig. 1.4, we can obtain

$$kx = \sin \beta \tag{1.30}$$

Fig. 1.3 Geometric interpretation of circle integrals

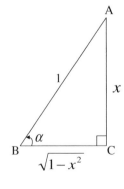

Fig. 1.4 Geometric interpretation of elliptic integrals

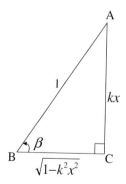

$$\sqrt{1 - k^2 x^2} = \cos \beta \tag{1.31}$$

$$\frac{kx}{\sqrt{1 - k^2 x^2}} = \tan \beta \tag{1.32}$$

Therefore, complete elliptic integral of the first kind $K(k)$ can be written as

$$K(k) = \int_0^{\frac{\pi}{2}} \frac{d\alpha}{\sqrt{1 - k^2 \sin^2 \alpha}} \tag{1.33}$$

Equation (1.33) can also be expressed as

$$K(k) = \int_0^{\sin^{-1} k} \frac{d\beta}{\sqrt{k^2 - \sin^2 \beta}} \tag{1.34}$$

With a generalization procedure similar to the above derivation, we can obtain ellipsoidal integral.

Definition Ellipsoid sine function $s_l(m, k, h)$ is defined as

$$m = \int_0^x \frac{dx}{\sqrt{1 - h^2 x^2} \sqrt{1 - k^2 x^2}} \tag{1.35}$$

where

$$0 \le k = \frac{\sqrt{a^2 - b^2}}{a} \le 1 \tag{1.36}$$

$$0 \leq h = \frac{\sqrt{a^2 - c^2}}{a} \leq 1 \tag{1.37}$$

Equations (1.36) and (1.37) give geometric parameters of the ellipsoid. Conventionally, we have

$$c \leq b \leq a \tag{1.38}$$

Hence we can get

$$h \geq k \tag{1.39}$$

$$\boxed{x = s_l(m, k, h)} \tag{1.40}$$

Equation (1.40) is denoted as the ellipsoid sine function. Accordingly, ellipsoid cosine h function can be expressed as

$$\boxed{e_l(m, k, h) = \sqrt{1 - h^2 x^2}} \tag{1.41}$$

Similarly, ellipsoid Δ function is denoted as

$$\boxed{d_l(m, k, h) = \sqrt{1 - k^2 x^2}} \tag{1.42}$$

Hence, we have the following identities as

$$\boxed{h^2 s_l^2(m, k, h) + e_l^2(m, k, h) \equiv 1} \tag{1.43}$$

and

$$\boxed{k^2 s_l^2(m, k, h) + d_l^2(m, k, h) \equiv 1} \tag{1.44}$$

In the ellipsoidal functions, the most prominent is that we do not define ellipsoid cosine function $c_l(m, k)$. There are two reasons as follows:

① According to $e_l(m, k, h)$, ellipsoid cosine h function can be reduced to ellipsoid cosine function when $h = 1$, e.g.,

$$e_l(m, k, 1) = c_l(m, k, 1) \tag{1.45}$$

	Circle integral	Elliptic integral	Ellipsoidal integral
Table 1.3 From the circle integrals to the ellipsoid integrals	$d = \int\limits_0^x \frac{dx}{\sqrt{1-x^2}}$	$u = \int\limits_0^x \frac{dx}{\sqrt{1-x^2}\sqrt{1-k^2x^2}}$	$m = \int\limits_0^x \frac{dx}{\sqrt{1-h^2x^2}\sqrt{1-k^2x^2}}$

② Function s_l, e_l, and d_l consist of a complete set. Note that in ellipsoid integrals (1.35) denominator of the integrand has only two factors $\sqrt{1 - h^2 x^2}$ and $\sqrt{1 - k^2 x^2}$. As an extra word, the definition of $c_l(m, k, 1)$ will bring unnecessary difficulties. Table 1.3 gives evolution from the circle integrals to the ellipsoid integrals.

Chapter 2
Application Background of Circle Integrals

On many occasions we repeatedly emphasize that application of engineering mathematics has the same importance as mathematics itself. Here we start the discussion from circle integral, e.g.,

$$\int_0^1 \frac{\mathrm{d}x}{\sqrt{1-x^2}} \tag{2.1}$$

Circle integral is related to *capacitance C of conducting circle disk*. We first consider a conducting circle disk with an infinitely small thickness shown in Fig. 2.1.

Assume charge density $\sigma(r)$ on the conducting circle disk. According to the symmetry of the disk, $\sigma(r)$ is irrelevant to azimuthal angle φ. Charge density on the periphery of the disk is distributed in the form of inverse square root, i.e.,

$$\sigma(r) = \frac{\sigma_0}{\sqrt{1 - \left(\frac{r}{a}\right)^2}} \tag{2.2}$$

in which σ_0 denotes a maximum magnitude of the charge density. As shown in Fig. 2.2, on the periphery of the disk the charge density has the inverse square root distribution.

The capacitance of the circle disk C is denoted as

$$\boxed{C = \frac{Q}{V}} \tag{2.3}$$

© Science Press, Beijing and Springer Nature Singapore Pte Ltd. 2017
C.-H. Liang, *Notes on the Ellipsoidal Function*,
DOI 10.1007/978-981-10-2908-0_2

Fig. 2.1 A conducting disk
with an infinitely small
thickness

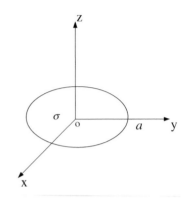

Fig. 2.2 The distribution of
the charge density over a
circle disk

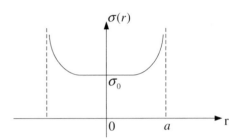

in which Q is total charge of the disk, and V is the electric potential. Hence we have

$$Q = \iint\limits_{s} \sigma(r) r \mathrm{d}r \mathrm{d}\varphi \tag{2.4}$$

$$V = \iint\limits_{s} \frac{\sigma(r)}{4\pi\varepsilon r} r \mathrm{d}r \mathrm{d}\varphi \tag{2.5}$$

Substituting Eq. (2.2) into Eq. (2.4), we have

$$
\begin{aligned}
Q &= \iint\limits_{s} \frac{\sigma_0}{\sqrt{1 - \left(\frac{r}{a}\right)^2}} r \mathrm{d}r \mathrm{d}\phi \\
&= 2\pi\sigma_0 a^2 \int_{0}^{a} \frac{-\frac{1}{2}\mathrm{d}\left[1 - \left(\frac{r}{a}\right)^2\right]}{\sqrt{1 - \left(\frac{r}{a}\right)^2}} \\
&= 2\pi\sigma_0 a^2 \frac{\frac{1}{2}\sqrt{1 - \left(\frac{r}{a}\right)^2}\Big|_{a}^{0}}{\frac{1}{2}} \\
&= 2\pi\sigma_0 a^2
\end{aligned}
\tag{2.6}
$$

Considering Eq. (2.6) and charge distribution shown in Fig. 2.2, we can completely think that the average charge density on the circle disk is

$$\bar{\sigma} = 2\sigma_0 \tag{2.7}$$

We rewrite Eq. (2.6) as

$$Q = \bar{\sigma}S = \bar{\sigma}\pi a^2 = 2\sigma_0\pi a^2 \tag{2.8}$$

On the other hand, the electric potential V can be expressed as

$$
\begin{aligned}
V &= \iint\limits_{s} \frac{\sigma_0 r \mathrm{d}r\mathrm{d}\phi}{4\pi\varepsilon r\sqrt{1 - \left(\frac{r}{a}\right)^2}}\\
&= 2\pi\sigma_0 a \int\limits_{0}^{a} \frac{\mathrm{d}\left(\frac{r}{a}\right)}{4\pi\varepsilon\sqrt{1 - \left(\frac{r}{a}\right)^2}}\\
&= \frac{2\pi\sigma_0 a}{4\pi\varepsilon} \sin^{-1}\left(\frac{r}{a}\right)\Big|_{0}^{a}\\
&= \frac{\pi\sigma_0 a}{4\varepsilon}
\end{aligned}
\tag{2.9}
$$

According to Eq. (2.3), we can obtain the capacitance as follows:

$$\boxed{C = \frac{Q}{V} = 8\varepsilon a} \tag{2.10}$$

Further, if we express the electric potential V in an integral form, we have

$$Q = 2\pi\sigma_0 a^2 \tag{2.11}$$

$$V = \frac{2\pi\sigma_0 a}{4\pi\varepsilon} \int\limits_{0}^{1} \frac{\mathrm{d}x}{\sqrt{1 - x^2}} \tag{2.12}$$

Therefore, we can get

$$\boxed{C = \frac{4\pi\varepsilon a}{\int_0^1 \frac{\mathrm{d}x}{\sqrt{1-x^2}}}} \tag{2.13}$$

It can be clearly seen that capacitance of the conducting circle disk is tightly related to the circle integral $\int_0^1 \mathrm{d}x \Big/ \sqrt{1 - x^2}$.

In addition, according to the above discussion, we can know that there are not circle integrals of the first and the second kinds. This is because factor $\sqrt{1 - x^2}$ is

always in the denominator of the studied problem. The second important application of the circle integral is to calculate the perimeter of the circle, i.e.,

$$L = 4a \int_0^1 \frac{dx}{\sqrt{1 - x^2}} = 4a\left(\frac{\pi}{2}\right) = 2\pi a. \tag{2.14}$$

Chapter 3
Application Background of Elliptic Integrals

Elliptic integral is an extension to circle integral. In this chapter, we start to study the practical application of the elliptic integral. Specifically, the elliptic integral is expressed as

$$\int_0^1 \frac{\mathrm{d}x}{\sqrt{(1-x^2)(1-k^2x^2)}} \tag{3.1}$$

Figure 3.1 shows a general ellipse with a semi-major axis a and a semi-minor axis b. Here we assume

$$a \geq b \tag{3.2}$$

and

$$k = \frac{\sqrt{a^2 - b^2}}{a} \tag{3.3}$$

We list two representative applications of two kinds of the elliptic integrals in Table 3.1.

3.1 Capacitance of Elliptic Conductor C

Here we use a unified conclusion given in the following application of the ellipsoidal integral to solve this problem. In the following chapter, we have given the proof for the capacitance of an arbitrary ellipsoid conductor as

© Science Press, Beijing and Springer Nature Singapore Pte Ltd. 2017
C.-H. Liang, *Notes on the Ellipsoidal Function*,
DOI 10.1007/978-981-10-2908-0_3

Fig. 3.1 Geometry of a
general ellipse

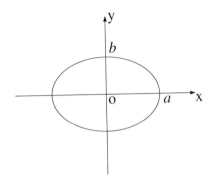

Table 3.1 Two representative applications of two kinds of elliptic integrals

	Elliptic integral of the first kind	Elliptic integral of the second kind
Definition	$K(k) = \int_0^1 \frac{dx}{\sqrt{1-x^2}\sqrt{1-k^2x^2}}$ $k = \frac{\sqrt{a^2-b^2}}{a}$	$E(k) = \int_0^1 \frac{\sqrt{1-k^2x^2}}{\sqrt{1-x^2}}dx$ $k = \frac{\sqrt{a^2-b^2}}{a}$
Representative applications	Capacitance C of conducting elliptic disk $C = 4\pi\varepsilon a / K(k) = 8\varepsilon a / \left[\frac{K(k)}{\pi/2}\right]$	Perimeter L of the ellipse $L = 4a \int_0^1 \sqrt{\frac{1-k^2x^2}{1-x^2}}dx = 4aE(k)$

$$C = \frac{4\pi\varepsilon a}{G(k,h)} \tag{3.4}$$

where the complete ellipsoidal integral of the first kind $G(k,h)$ is denoted as

$$G(k,h) = \int_0^1 \frac{dx}{\sqrt{1-h^2x^2}\sqrt{1-k^2x^2}} \tag{3.5}$$

and

$$k = \frac{\sqrt{a^2-b^2}}{a}, \quad h = \frac{\sqrt{a^2-c^2}}{a} \tag{3.6}$$

The capacitance given by Eq. (3.4) can be reduced to the case of the conducting elliptic disk with

$$c = 0, \quad h = 1 \qquad (3.7)$$

In this case, we can obtain

$$K(k) = G(k, 1) = \int_0^1 \frac{dx}{\sqrt{1 - x^2}\sqrt{1 - k^2 x^2}} \qquad (3.8)$$

Especially when $k = 0$, we have

$$K(0) = \int_0^1 \frac{dx}{\sqrt{1 - x^2}} = \sin^{-1}(1) = \frac{\pi}{2} \qquad (3.9)$$

which corresponds to the circle integral. Hence we can write the capacitance of the conducting elliptic disk in a general form as

$$C = \frac{8\varepsilon a}{\left[\frac{K(k)}{\pi/2}\right]} \qquad (3.10)$$

Especially for the case of the conducting circle disk with $a = b$, considering Eq. (3.9) we can further obtain capacitance of the conducting circle disk as follows

$$C = 8\varepsilon a \qquad (3.11)$$

3.2 Perimeter of Ellipse L

Length of an arbitrary plane curve can be expressed in the form of the integral as

$$L = \int dl = \int \sqrt{1 + \left(\frac{dy}{dx}\right)^2}\, dx \qquad (3.12)$$

Considering the symmetry of the ellipse, we only need to solve the corresponding part in the first quadrant with a length of $L/4$, as shown in Fig. 3.2.

The equation of the ellipse in the first quadrant can be described as

$$\frac{x^2}{a^2} + \frac{y^2}{b^2} = 1, \quad x \geq 0, \ y \geq 0$$

$$a \geq b$$

Fig. 3.2 The length of the
ellipse in the first quadrant

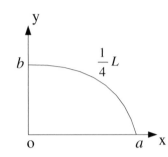

And then we have

$$y = b\sqrt{1 - \left(\frac{x}{a}\right)^2} \tag{3.13}$$

According to Eq. (3.13), we can easily obtain

$$\frac{dy}{dx} = -\frac{\left(\frac{b}{a}\right)\left(\frac{x}{a}\right)}{\sqrt{1 - \left(\frac{x}{a}\right)^2}} \tag{3.14}$$

Hence we have

$$\left(\frac{dy}{dx}\right)^2 = \frac{\left(\frac{b}{a}\right)^2\left(\frac{x}{a}\right)^2}{1 - \left(\frac{x}{a}\right)^2} \tag{3.15}$$

From Eq. (3.15), we can get

$$1 + \left(\frac{dy}{dx}\right)^2 = \frac{1 - k^2\left(\frac{x}{a}\right)^2}{1 - \left(\frac{x}{a}\right)^2} \tag{3.16}$$

Assuming

$$t = \left(\frac{x}{a}\right), \quad t \in [0, 1] \tag{3.17}$$

Finally we can obtain

$$L = 4a \int_0^1 \sqrt{\frac{1 - k^2 t^2}{1 - t^2}} dt = 4aE(k) \tag{3.18}$$

in which the complete elliptic integral of the second kind is

$$E(k) = \int_0^1 \sqrt{\frac{1 - k^2 t^2}{1 - t^2}} dt \tag{3.19}$$

Considering the case of $k = 0$, Eq. (3.19) can be reduced to

$$E(0) = \int_0^1 \sqrt{\frac{1}{1 - t^2}} dt = \sin^{-1}(1) = \frac{\pi}{2} \tag{3.20}$$

which corresponds to the perimeter of the circle. Specifically, we have

$$L = 4aE(0) = 2\pi a \tag{3.21}$$

Now we review Table 1.1 in the Chap. 1. It can be seen that due to the introduction of complete elliptic integral of the second kind, the perimeter of the ellipse can be solved.

According to the above derivation, we can also find out that

$$K(0) = E(0) = \frac{\pi}{2}. \tag{3.22}$$

Chapter 4
Application Background of Ellipsoidal Integral

Figure 4.1 presents the general geometrical structure of an ellipsoid, with the equation being

$$\frac{x^2}{a^2} + \frac{y^2}{b^2} + \frac{z^2}{c^2} = 1$$

where a, b, and c are three semi-axes separately. Without loss of generality, assume $a \geq b \geq c$.

Ellipsoidal integral is the further extension of elliptic integral. Let us start studying the actual application of the following ellipsoidal integral, namely

$$\int_0^1 \frac{\mathrm{d}x}{\sqrt{1 - h^2 x^2}\sqrt{1 - k^2 x^2}}$$

where

$$k = \frac{\sqrt{a^2 - b^2}}{a}, \quad h = \frac{\sqrt{a^2 - c^2}}{a}$$

Table 4.1 lists the typical applications of complete ellipsoidal integrals of the two kinds. We can see from Table 4.1 that in the case of the ellipsoid, complete ellipsoidal integrals of the first kind are not symmetrical with that of the second kind.

In the definition of complete ellipsoidal integral of the second kind $H(e, h)$, the integrand is

© Science Press, Beijing and Springer Nature Singapore Pte Ltd. 2017
C.-H. Liang, *Notes on the Ellipsoidal Function*,
DOI 10.1007/978-981-10-2908-0_4

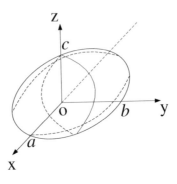

Fig. 4.1 General geometrical structure of an ellipsoid

Table 4.1 Typical applications of complete ellipsoidal integrals of the two kinds

	Complete ellipsoidal integral of the first kind	Complete ellipsoidal integral of the second kind
Definition	$G(k,h) = \int_0^1 \frac{\mathrm{d}x}{\sqrt{1-h^2x^2}\sqrt{1-k^2x^2}}$ $k = \frac{\sqrt{a^2-b^2}}{a}, \; h = \frac{\sqrt{a^2-c^2}}{a}$	$H(e,h) = \int_0^1 \sqrt{1-h^2x^2}\, E\left(\frac{e\sqrt{1-x^2}}{\sqrt{1-h^2x^2}}\right)\mathrm{d}x$ $e = \frac{\sqrt{b^2-c^2}}{b}, \; h = \frac{\sqrt{a^2-c^2}}{a}$
Typical application	Capacitance of an arbitrary ellipsoid conductor $C = \frac{4\pi\varepsilon a}{G(k,h)}$	Surface area of an arbitrary ellipsoid $S = 8abH(e,h)$

$$E(\bar{k}) = \int\limits_0^1 \sqrt{\frac{1-\bar{k}^2t^2}{1-t^2}}\,\mathrm{d}t$$

where

$$\bar{k} = \frac{e\sqrt{1-x^2}}{\sqrt{1-h^2x^2}}, \quad e = \frac{\sqrt{b^2-c^2}}{b}$$

Meanwhile, Table 4.1 also shows that complete ellipsoidal integrals of the two kinds have two independent ellipsoidal parameters.

The last and most important point is as follows: Although elliptic integrals deal with problems of two-dimensional ellipses, they are defined as one-dimensional integrals; and ellipsoidal integrals deal with problems of three-dimensional ellipsoids, but they are defined as one-dimensional integrals again. This is really interesting and worthy of further exploration.

4.1 Capacitance of Ellipsoid Conductor C

It is extremely difficult to calculate the capacitance C of an ellipsoid conductor. To our knowledge, currently there are three special and relatively simple cases derived by famous experts in electromagnetics such as J.A. Stratton. The following is a brief introduction to their study findings.

Case 1 Oblate Ellipsoid of Revolution

Figure 4.2 presents the geometrical structure of the oblate ellipsoid of revolution.

The case is equivalent to parameter $k = \frac{\sqrt{a^2-b^2}}{a} = 0$. Stratton derived

$$C = \frac{4\pi\varepsilon a}{\left[\frac{\tan^{-1}\left(h/\sqrt{1-h^2}\right)}{h}\right]} \tag{4.1}$$

Figure 4.3 delineates basic geometrical relationship of the triangle h. It can be seen easily from Fig. 4.3 that

$$\alpha = \tan^{-1}\left(\frac{h}{\sqrt{1-h^2}}\right) = \sin^{-1}(h) \tag{4.2}$$

Therefore, C in Eq. (4.1) can be rewritten as

Fig. 4.2 Oblate ellipsoid of revolution $(a = b \geq c)$

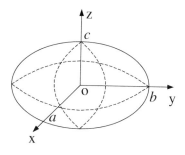

Fig. 4.3 Basic geometrical
relationship of index ellipsoid
$h = \frac{\sqrt{a^2 - c^2}}{a}$

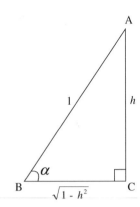

$$C = \frac{4\pi\varepsilon a}{\left[\frac{\sin^{-1}(h)}{h}\right]} \qquad (4.3)$$

When $h \to 0$, it tends to be a sphere. Considering

$$\lim_{h \to 0} \left[\frac{\sin^{-1}(h)}{h}\right] = 1 \qquad (4.4)$$

the capacitance C_0 of the sphere can be obtained according to Eq. (4.3) as

$$C_0 = 4\pi\varepsilon a \qquad (4.5)$$

Case 2 Prolate ellipsoid of revolution

Figure 4.4 presents the geometrical structure of the prolate ellipsoid of revolution.

The case is corresponding to

$$k = h \qquad (4.6)$$

Fig. 4.4 Prolate ellipsoid of
revolution $(a \geq b = c)$

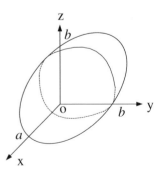

The derived capacitance C is

$$C = \frac{4\pi\varepsilon a}{\left[\frac{\frac{1}{2}\ln\left(\frac{1+h}{1-h}\right)}{h}\right]} = \frac{4\pi\varepsilon a}{\left[\frac{th^{-1}(h)}{h}\right]} \tag{4.7}$$

With the completely same discussion idea as Eq. (4.3), we have

$$\lim_{h\to 0}\left[\frac{th^{-1}(h)}{h}\right] = 1 \tag{4.8}$$

We again obtain the capacitance of the sphere conductor

$$C_0 = 4\pi\varepsilon a$$

Case 3 Elliptical conductor disk

Figure 4.5 presents the geometrical structure of the elliptical conductor disk.

Elliptical conductor disk is essentially a two-dimensional problem. However, here it is taken as a limit case of three-dimensional ellipsoid ($c = 0$), or as an example with known results, namely

$$C = \frac{4\pi\varepsilon a}{K(k)} \tag{4.9}$$

Table 4.2 gives the three known results of capacitance of ellipsoid conductor C.

Fig. 4.5 Elliptical conductor disk ($c = 0$)

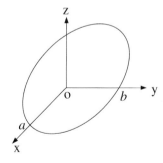

Table 4.2 Three known results of the capacitance of ellipsoid conductor C

Oblate ellipsoid of revolution $a = b \geq c$	$C = \dfrac{4\pi\varepsilon a}{\left[\frac{\sin^{-1}(h)}{h}\right]}$
Prolate ellipsoid of revolution $a \geq b = c$ $k = h$	$C = \dfrac{4\pi\varepsilon a}{\left[\frac{th^{-1}(h)}{h}\right]}$
Elliptical conductor disk $c = 0$ $h = 1$	$C = \dfrac{4\pi\varepsilon a}{K(k)} = \dfrac{4\pi\varepsilon a}{\int_0^1 \frac{dx}{\sqrt{(1-x^2)(1-k^2x^2)}}}$

4.2   Complete Ellipsoidal Integral of the First Kind

The evolvement of many theories usually goes through three stages, namely conjecture, testing and proving. Table 4.3 presents the conjecture process of complete ellipsoidal integral of the first kind.

Table 4.3 shows that the conjecture is simple and clear, because when

$$h = 1 \tag{4.10}$$

the result is degraded into the capacitance of elliptical conductor disk ($c = 0$). Of course, conjecture cannot replace proof, but the process cannot be left out.

Case 1 Oblate ellipsoid of revolution ($a = b \geq c$)

It is expounded above that the case is equivalent to $k = 0$. We introduce

$$y = hx \tag{4.11}$$

If we conjecture

$$G(k, h) = \int_0^1 \frac{dx}{\sqrt{1 - h^2 x^2} \sqrt{1 - k^2 x^2}} \tag{4.12}$$

where $k = \frac{\sqrt{a^2 - b^2}}{a}, h = \frac{\sqrt{a^2 - c^2}}{a}$. The conjectural expression of capacitance can be given as

$$C = \frac{4\pi \varepsilon a}{G(k, h)} \tag{4.13}$$

In this case, $G(k, h)$ can be rewritten as

$$G(k, h) = \frac{1}{h} \int_0^h \frac{dy}{\sqrt{1 - y^2} \sqrt{1 - \left(\frac{k}{h}\right)^2 y^2}} \tag{4.14}$$

Table 4.3 Conjecture of the capacitance of ellipsoid conductor C

	Capacitance C	Integral expression
Elliptical conductor disk	$C = \frac{4\pi \varepsilon a}{K(k)}$	$K(k) = \int_0^1 \frac{dx}{\sqrt{1 - x^2}\sqrt{1 - k^2 x^2}}$
Ellipsoid conductor	$C = \frac{4\pi \varepsilon a}{G(k,h)}$	$G(k, h) = \int_0^1 \frac{dx}{\sqrt{1 - h^2 x^2}\sqrt{1 - k^2 x^2}}$

Particularly when $k = 0$, we can further obtain

$$G(0, h) = \frac{1}{h} \int_0^h \frac{dy}{\sqrt{1 - y^2}} = \frac{1}{h} \sin^{-1}(h) \tag{4.15}$$

which is completely identical to the first result in Table 4.2. Namely, the capacitance of the oblate ellipsoid of revolution can be expressed as

$$C = \frac{4\pi\varepsilon a}{\left[\frac{\sin^{-1}(h)}{h}\right]} = \frac{4\pi\varepsilon a}{G(0, h)} \tag{4.16}$$

The first step in conjecture is successful.

Case 2 Prolate ellipsoid of revolution $(a \geq b = c)$
   The case is equivalent to $k = h$. Suppose $y = hx$ again, whereupon we get

$$G(h, h) = \int_0^1 \frac{dx}{1 - h^2 x^2} = \frac{1}{h} \int_0^h \frac{dy}{1 - y^2} = \frac{1}{2h} \ln\left(\frac{1 + h}{1 - h}\right) = \frac{th^{-1}(h)}{h} \tag{4.17}$$

We can obtain the capacitance of the prolate ellipsoid of revolution through comparison with the second result in Table 4.2.

$$C = \frac{4\pi\varepsilon a}{\frac{1}{2h} \ln\left(\frac{1 + h}{1 - h}\right)} = \frac{4\pi\varepsilon a}{\left[\frac{th^{-1}(h)}{h}\right]} = \frac{4\pi\varepsilon a}{G(h, h)} \tag{4.18}$$

We have another success. As the incidental result, we have

$$G(0, 0) \equiv 1 \tag{4.19}$$

which is the case of degradation to sphere.

Case 3 Elliptical conductor disk
   This case happens to be the starting point of conjecture; it needs no repetition here. It is clear that the cases above (merely three) cannot replace proof. We employ the numerical method—Method of Moments (MoM) to calculate them and compare the results with $G(k, h)$, which are shown in Tables 4.4, 4.5 and 4.6.
   We offer the precise theorem in the following section.

Theorem *The capacitance of any ellipsoid conductor can be expressed as*

$$C = \frac{4\pi\varepsilon a}{G(k, h)} \tag{4.20}$$

Table 4.4 Result c/ε calculated through $c = 4\pi\varepsilon a/G(k,h)$, $(a = 1)$

c \ b	1.0	0.9	0.8	0.7	0.6	0.5	0.4	0.3	0.2	0.1
1.0	12.56637									
0.9	12.14463	11.72559								
0.8	11.71687	11.30042	10.87766							
0.7	11.28262	10.86860	10.44807	10.02044						
0.6	10.84131	10.42956	10.01101	9.58499	9.15072					
0.5	10.39230	9.9826	9.56572	9.14091	8.70724	8.26359				
0.4	9.93485	9.52693	9.11134	8.6872	8.25339	7.80853	7.35083			
0.3	9.46807	9.06159	8.64681	8.22264	7.78772	7.34026	6.87791	6.39739		
0.2	8.9909	8.58543	8.17083	7.74572	7.30836	6.85638	6.3865	5.89389	5.37092	
0.1	8.5026	8.09706	7.68180	7.25453	6.81289	6.3536	5.87182	5.35986	4.80427	4.17723
0.0	8	7.59473	7.17769	6.74563	6.29798	5.82716	5.3264	4.78214	4.16641	3.40032

Table 4.5 Result c/ε calculated through MoM

c	b									
	1.0	0.9	0.8	0.7	0.6	0.5	0.4	0.3	0.2	0.1
1.0	12.5560									
0.9	12.1362	11.7187								
0.8	11.7080	11.2894	10.8668							
0.7	11.2725	10.8613	10.4386	10.0114						
0.6	10.8281	10.4186	10.0016	9.5764	9.1420					
0.5	10.3773	9.9711	9.5561	9.1324	8.6993	8.2567				
0.4	9.9219	9.5194	9.1003	8.6784	8.2460	7.8016	7.3448			
0.3	9.4559	9.0477	8.6379	8.2137	7.7783	7.3328	6.8698	6.3922		
0.2	8.9784	8.5759	8.1597	7.7378	7.3011	6.8481	6.3788	5.8895	5.3655	
0.1	8.4918	8.0859	7.6659	7.2409	6.8066	6.3481	5.8665	5.3551	4.7983	4.1703
0.0	7.9719	7.5692	7.1565	6.7254	6.2805	5.8187	5.3165	4.7740	4.1574	3.3900

where

$$G(k,h) = \int_0^1 \frac{dx}{\sqrt{1-h^2x^2}\sqrt{1-k^2x^2}}$$

$$k = \frac{\sqrt{a^2-b^2}}{a}, \quad h = \frac{\sqrt{a^2-c^2}}{a} \tag{4.21}$$

$$a \geq b \geq c$$

Prove Stratton, in his famous *Electromagnetic Theory*, offered the Laplace equation under the ellipsoidal coordinates (u, v, w) as

$$(v-u)R_u \frac{\partial}{\partial u}\left(R_u \frac{\partial \varphi}{\partial u}\right) + (w-u)R_v \frac{\partial}{\partial v}\left(R_v \frac{\partial \varphi}{\partial v}\right) + (u-v)R_w \frac{\partial}{\partial w}\left(R_w \frac{\partial \varphi}{\partial w}\right) = 0 \tag{4.22}$$

Stratton pointed out that the solution φ to Eq. (4.22) on the surface of the ellipsoid of

$$u = 0 \tag{4.23}$$

is independent of v and w, and behaves inversely proportional to distance r at infinity, being expressed as

$$\varphi = \varphi(u) \tag{4.24}$$

Table 4.6 Relative error σ between the two methods

c \ b	1.0	0.9	0.8	0.7	0.6	0.5	0.4	0.3	0.2	0.1
1.0	8.259 E−4									
0.9	6.496 E−4	9.761 E−4								
0.8	7.576 E−4	6.721 E−4	9.994 E−4							
0.7	8.978 E−4	0.00105	9.072 E−4	9.030 E−4						
0.6	0.00122	0.00115	9.408 E−4	8.970 E−4	9.538 E−4					
0.5	0.00145	7.910 E−4	0.00101	9.318 E−4	9.127 E−4	8.345 E−4				
0.4	0.00131	0.00154	0.00121	0.00101	8.962 E−4	8.883 e−4	8.210 E−4			
0.3	0.00129	0.00111	0.00103	0.00109	0.00121	0.00102	0.00118	8.119 E−4		
0.2	0.00139	0.00138	0.00136	0.00102	9.944 E−4	0.00121	0.00121	7.945 E−4	0.00124	
0.1	0.00121	0.00337	0.00207	0.00188	9.241 E−4	8.664 E−4	9.068 E−4	8.889 E−4	0.00217	0.00166
0.0	0.00352	9.761 E−4	0.00296	0.00314	0.00278	0.00145	0.00186	0.00171	0.00101	0.00304

Equation (4.22) is simplified as

$$\frac{\partial}{\partial u}\left(R_u \frac{\partial \varphi}{\partial u}\right) = 0 \qquad (4.25)$$

where

$$R_u = \sqrt{(u+a^2)(u+b^2)(u+c^2)} \qquad (4.26)$$

The ellipsoid potential $\varphi(u)$ derived by Stratton satisfies

$$\boxed{\varphi(u) = \frac{Q}{8\pi\varepsilon}\int\limits_u^\infty \frac{du}{R_u}} \qquad (4.27)$$

in which Q denotes the total electric charge on the surface of the ellipsoid conductor. Therefore, in the theorem we only need to prove the following: with the potential $\varphi(u)$ offered by Eq. (4.27), the capacitance of the ellipsoid conductor is

$$\begin{cases} C = \frac{4\pi\varepsilon a}{G(k,h)} \\ G(k,h) = \int\limits_0^1 \frac{dx}{\sqrt{1-h^2x^2}\sqrt{1-k^2x^2}} \end{cases}$$

We notice that

$$\varphi(0) = \int\limits_0^\infty \frac{du}{R_u} \cdot \frac{Q}{8\pi\varepsilon} = \frac{Q}{8\pi\varepsilon}\int\limits_0^\infty \frac{du}{\sqrt{(u+a^2)(u+b^2)(u+c^2)}}$$

$$= \frac{Q}{4\pi\varepsilon}\int\limits_0^\infty \frac{d(\sqrt{u+a^2})}{\sqrt{(u+b^2)(u+c^2)}} \qquad (4.28)$$

Reintroducing the parameters

$$s = \sqrt{u+a^2} \qquad (4.29)$$

we know that

$$u = s^2 - a^2 \qquad (4.30)$$

We again express Eq. (4.28) as

$$\varphi(0) = \frac{Q}{4\pi\varepsilon} \int_0^\infty \frac{\mathrm{d}s}{\sqrt{[s^2 - (a^2 - b^2)][s^2 - (a^2 - c^2)]}} \tag{4.31}$$

In this case, we further introduce

$$x = \frac{a}{s} \tag{4.32}$$

$$\mathrm{d}s = -a\frac{\mathrm{d}x}{x^2} \tag{4.33}$$

Observing Eq. (4.31) again, we obtain

$$\begin{aligned}
\varphi(0) &= \frac{Q}{4\pi\varepsilon} \int_0^1 \frac{a\mathrm{d}x}{x^2\sqrt{\left[\left(\frac{a}{x}\right)^2 - (a^2 - b^2)\right]\left[\left(\frac{a}{x}\right)^2 - (a^2 - c^2)\right]}} \\
&= \frac{Q}{4\pi\varepsilon a} \int_0^1 \frac{\mathrm{d}x}{\sqrt{[1 - h^2 x^2][1 - k^2 x^2]}} \\
&= \frac{Q}{4\pi\varepsilon a} G(k, h)
\end{aligned} \tag{4.34}$$

Finally, we write the capacitance of the ellipsoid conductor as

$$C = \frac{Q}{\varphi(0)} = \frac{4\pi\varepsilon a}{G(k, h)} \tag{4.35}$$

$G(k, h)$ is hence called complete elliptic integral of the first kind, and

$$G(k, h) = \int_0^1 \frac{\mathrm{d}x}{\sqrt{[1 - h^2 x^2][1 - k^2 x^2]}} \tag{4.36}$$

Following the above procedure, it clearly shows that many special functions are proposed simply for the purpose of solving practical problems.

4.3 Complete Ellipsoidal Integral of the Second Kind

The idea behind further questioning is clear: Now that complete elliptic integral of the second kind $E(k)$ corresponds to the perimeter L of the ellipse, could the surface area S of the ellipsoid produce complete ellipsoidal integral of the second kind? Figure 4.6 presents the surface area S of the ellipsoid.

Fig. 4.6 Surface area S of an ellipsoid

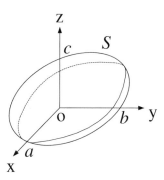

Conjecture stems from practice, which does deserve exploration. The ellipsoid equation is

$$\frac{x^2}{a^2} + \frac{y^2}{b^2} + \frac{z^2}{c^2} = 1 \tag{4.37}$$

First, we give the general expression of curved surface differential element as

$$ds = \sqrt{1 + \left(\frac{\partial z}{\partial x}\right)^2 + \left(\frac{\partial z}{\partial y}\right)^2}\, dxdy \tag{4.38}$$

The surface area of the ellipsoid in the first quadrant is calculated, and the total surface area S is eight times that of the ellipsoid in the first quadrant, namely

$$S = 8 \iint\limits_{S} \sqrt{1 + \left(\frac{\partial z}{\partial x}\right)^2 + \left(\frac{\partial z}{\partial y}\right)^2}\, dxdy \tag{4.39}$$

The biggest advantage here is that x, y, and z in the first quadrant are not negative. Without loss of generality, assuming $a \geq b \geq c$, we have

$$z = c\sqrt{1 - \left[\left(\frac{x}{a}\right)^2 + \left(\frac{y}{b}\right)^2\right]} \tag{4.40}$$

We can get from Eq. (4.40)

$$\left(\frac{\partial z}{\partial x}\right) = -\frac{\left(\frac{c}{a}\right)\left(\frac{x}{a}\right)}{\sqrt{1 - \left[\left(\frac{x}{a}\right)^2 + \left(\frac{y}{b}\right)^2\right]}} \tag{4.41}$$

$$\left(\frac{\partial z}{\partial y}\right) = -\frac{\left(\frac{c}{b}\right)\left(\frac{y}{b}\right)}{\sqrt{1 - \left[\left(\frac{x}{a}\right)^2 + \left(\frac{y}{b}\right)^2\right]}} \tag{4.42}$$

and

$$\left(\frac{\partial z}{\partial x}\right)^2 = \frac{\left(\frac{c}{a}\right)^2\left(\frac{x}{a}\right)^2}{1 - \left[\left(\frac{x}{a}\right)^2 + \left(\frac{y}{b}\right)^2\right]} \tag{4.43}$$

$$\left(\frac{\partial z}{\partial y}\right)^2 = \frac{\left(\frac{c}{b}\right)^2\left(\frac{y}{b}\right)^2}{1 - \left[\left(\frac{x}{a}\right)^2 + \left(\frac{y}{b}\right)^2\right]} \tag{4.44}$$

Combining Eqs. (4.43) and (4.44), we obtain

$$1 + \left(\frac{\partial z}{\partial x}\right)^2 + \left(\frac{\partial z}{\partial y}\right)^2 = \frac{1 - \left[\left(\frac{x}{a}\right)^2 + \left(\frac{y}{b}\right)^2\right] + \left(\frac{c}{a}\right)^2\left(\frac{x}{a}\right)^2 + \left(\frac{c}{b}\right)^2\left(\frac{y}{b}\right)^2}{1 - \left[\left(\frac{x}{a}\right)^2 + \left(\frac{y}{b}\right)^2\right]} \tag{4.45}$$

Equation (4.45) can be simplified into

$$1 + \left(\frac{\partial z}{\partial x}\right)^2 + \left(\frac{\partial z}{\partial y}\right)^2 = \frac{1 - h^2\left(\frac{x}{a}\right)^2 - e^2\left(\frac{y}{b}\right)^2}{1 - \left[\left(\frac{x}{a}\right)^2 + \left(\frac{y}{b}\right)^2\right]} \tag{4.46}$$

$$h = \frac{\sqrt{a^2 - c^2}}{a} \tag{4.47}$$

$$e = \frac{\sqrt{b^2 - c^2}}{b} \tag{4.48}$$

Substituting Eq. (4.46) into Eq. (4.39), we have

$$S = 8 \iint_s \sqrt{\frac{1 - h^2\left(\frac{x}{a}\right)^2 - e^2\left(\frac{y}{b}\right)^2}{1 - \left(\frac{x}{a}\right)^2 - \left(\frac{y}{b}\right)^2}} \, dxdy \tag{4.49}$$

Introduce new parameters

$$x' = \frac{x}{a} \tag{4.50}$$

$$y' = \frac{y}{b} \tag{4.51}$$

Fig. 4.7 1/4 unit circle in the
first quadrant of $x'o'y'$ plane

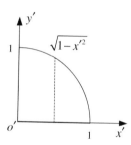

In the first quadrant of $x'o'y'$ plane, the parameters x' and y' describe one part of the unit circle, shown as Fig. 4.7, where $y' \in [0, \sqrt{1 - x'^2}]$ and $x' \in [0, 1]$.

Again, we express Eq. (4.49) using new parameters as

$$S = 8ab \int_0^1 dx' \int_0^{\sqrt{1-x'^2}} \sqrt{\frac{1 - h^2 x'^2 - e^2 y'^2}{1 - x'^2 - y'^2}} dy' \tag{4.52}$$

First, we study the integral with respect to y' in Eq. (4.52), namely

$$\int_0^{\sqrt{1-x'^2}} \sqrt{\frac{1 - h^2 x'^2 - e^2 y'^2}{1 - x'^2 - y'^2}} dy' \tag{4.53}$$

We rewrite the integrant in the equation as

$$\sqrt{\frac{1 - h^2 x'^2 - e^2 y'^2}{1 - x'^2 - y'^2}} = \sqrt{\frac{1 - h^2 x'^2}{1 - x'^2}} \frac{\sqrt{1 - \frac{e^2 y'^2}{1 - h^2 x'^2}}}{\sqrt{1 - \frac{y'^2}{1 - x'^2}}} \tag{4.54}$$

Furthermore, introduce the new parameter y'' as

$$\begin{cases} y'' = \frac{y'}{\sqrt{1-x'^2}} \\ dy'' = \frac{dy'}{\sqrt{1-x'^2}} \\ y'' \in [0, 1] \end{cases} \tag{4.55}$$

The ellipsoid surface integral can be expressed as

$$S = 8ab \int_0^1 \sqrt{1 - h^2 x'^2} dx' \int_0^1 \sqrt{\frac{1 - \bar{k}^2 y''^2}{1 - y''^2}} dy'' \tag{4.56}$$

where

$$\bar{k} = \frac{e\sqrt{1-x'^2}}{\sqrt{1-h^2x'^2}} \tag{4.57}$$

It is clear that the integral with respect to y'' fully matches the complete elliptic integral of the second kind, i.e.,

$$E(\bar{k}) = \int_0^1 \sqrt{\frac{1-\bar{k}^2y''^2}{1-y''^2}} \, dy'' = E\left(\frac{e\sqrt{1-x'^2}}{\sqrt{1-h^2x'^2}}\right) \tag{4.58}$$

The surface area of the ellipsoid can be expressed as

$$S = 8ab \int_0^1 \sqrt{1-h^2x'^2} \, E\left(\frac{e\sqrt{1-x'^2}}{\sqrt{1-h^2x'^2}}\right) dx' \tag{4.59}$$

Thus, we can introduce complete ellipsoidal integral of the second kind from the study of the surface area of the ellipsoid, which is

$$H(e,h) = \int_0^1 \sqrt{1-h^2t^2} \, E\left(\frac{e\sqrt{1-t^2}}{\sqrt{1-h^2t^2}}\right) dt \tag{4.60}$$

where

$$\begin{cases} e = \frac{\sqrt{b^2-c^2}}{b} \\ h = \frac{\sqrt{a^2-c^2}}{a} \end{cases}$$

It should be mentioned that the parameter t adopted in the above definition does not affect the essence of complete elliptic integral of the second kind (definite integral is irrelevant to the expression of parameter x' or t). Again, the following should be emphasized: Complete ellipsoidal integral of the second kind is one-dimensional too, but it includes two independent ellipsoidal parameters e and h. Finally, the general equation of the surface area S of the ellipsoid can be expressed as

$$S = 8abH(e,h) \tag{4.61}$$

We are happy to see that the problem of the ellipsoid characteristic quantity is fully resolved by the introduction of complete elliptic integral of the second kind. Additionally, the first and the second kinds defined with elliptic integrals are in symmetry, while the first and the second kinds in the ellipsoidal integrals differ from each other in expression. There are so many mysteries for us to explore.

Let us start the discussion on converting general cases into special cases.

Case 1 Sphere

The conditions are

$$a = b = c = R \tag{4.62}$$

The ellipsoidal parameters are

$$e = h = 0 \tag{4.63}$$

The integrand in the complete elliptic integral of the second kind is expressed according to Eq. (4.60) as

$$E\left(\frac{e\sqrt{1 - t^2}}{\sqrt{1 - h^2 t^2}}\right) = E(0) = \int_0^1 \frac{dx}{\sqrt{1 - x^2}} = \sin^{-1}(1) = \frac{\pi}{2} \tag{4.64}$$

$$\sqrt{1 - h^2 t^2} = 1 \tag{4.65}$$

Hence we have

$$H(0, 0) = \frac{\pi}{2} \tag{4.66}$$

Finally, we obtain the surface area S_0 of the sphere

$$S_0 = 8abH(0, 0) = 8ab \cdot \left(\frac{\pi}{2}\right) = 4\pi R^2 \tag{4.67}$$

Case 2 Elliptical conductor disk

The concrete conditions are

$$c = 0 \tag{4.68}$$

The ellipsoidal parameters deduced are

$$e = h = 1 \tag{4.69}$$

Reviewing Eq. (4.56), we obtain

$$\bar{k} = 1 \tag{4.70}$$

Hence we get

$$E(1) = 1 \tag{4.71}$$

Substituting it into the ellipsoid surface integral equation, we have

$$S = 8ab \int_0^1 \sqrt{1 - x'^2} \, dx' \tag{4.72}$$

Assuming

$$\begin{cases} x' = \sin\varphi \\ dx' = \cos\varphi \, d\varphi \\ \sqrt{1 - x'^2} = \cos\varphi \\ \varphi \in [0, \frac{\pi}{2}] \end{cases} \tag{4.73}$$

and substituting Eq. (4.73) into Eq. (4.72), we get

$$\begin{aligned} S &= 8ab \int_0^{\frac{\pi}{2}} \cos^2\varphi \, d\varphi \\ &= 8ab \int_0^{\frac{\pi}{2}} \frac{1}{2}(1 + \cos 2\varphi) \, d\varphi \\ &= 8ab \cdot \left(\frac{\pi}{4}\right) \end{aligned} \tag{4.74}$$

Finally, we obtain the surface area of the elliptical conductor disk

$$S = 2\pi ab \tag{4.75}$$

It is noted that the area of the elliptical conductor disk should be πab. When the third axis (z-axis) is considered, the limit case corresponding to $c = 0$ is the area $2\pi ab$ of the elliptical conductor disk of both sides, namely the results offered by Eq. (4.75).

Case 3 Oblate ellipsoid of revolution
The concrete conditions are

$$a = b \geq c \tag{4.76}$$

The corresponding ellipsoidal parameters are

$$e = h \tag{4.77}$$

And the surface area of the ellipsoid is

$$S = 8ab \int_0^1 \sqrt{1 - h^2 x^2} E\left(\frac{h\sqrt{1 - x^2}}{\sqrt{1 - h^2 x^2}}\right) dx \tag{4.78}$$

$$= 8abH(h, h)$$

No explicit expression can be offered for the case.

Case 4 Prolate ellipsoid of revolution
In this case, we have

$$a \geq b = c \tag{4.79}$$

The corresponding ellipsoidal parameters are

$$\begin{cases} e = 0 \\ h = k = \frac{\sqrt{a^2 - b^2}}{a} \end{cases} \tag{4.80}$$

We notice

$$E(0) = \int_0^1 \frac{dx}{\sqrt{1 - x^2}} = \frac{\pi}{2} \tag{4.81}$$

and

$$H(0, k) = \left(\frac{\pi}{2}\right) \int_0^1 \sqrt{1 - k^2 x^2} dx \tag{4.82}$$

Assuming

$$\begin{cases} t = kx \\ dt = kdx \\ t \in [0, k] \end{cases} \tag{4.83}$$

we have

$$H(0, k) = \frac{\pi}{2k} \int_0^k \sqrt{1 - t^2} dt \tag{4.84}$$

We assume again

$$\begin{cases} t = \sin\alpha \\ \mathrm{d}t = \cos\alpha\mathrm{d}\alpha \\ \sqrt{1-t^2} = \cos\alpha \\ t \in [0, \sin^{-1}k] \end{cases} \tag{4.85}$$

Hence, we can obtain

$$\int_0^k \sqrt{1-t^2}\mathrm{d}t = \int_0^{\sin^{-1}(k)} \cos^2\alpha\mathrm{d}\alpha$$

$$= \frac{1}{2}\int_0^{\sin^{-1}(k)} (1 + \cos 2\alpha)\mathrm{d}\alpha \tag{4.86}$$

$$= \frac{1}{2}\left(\alpha + \frac{1}{2}\sin 2\alpha\right)$$

$$= \frac{1}{2}(\alpha + \sin\alpha\cos\alpha)\Big|_0^{\sin^{-1}(k)}$$

$$= \frac{1}{2}\left(\sin^{-1}(k) + k\sqrt{1-k^2}\right)$$

Substituting Eq. (4.86) into Eq. (4.84), we have

$$H(0,k) = \frac{\pi}{4k}[\sin^{-1}(k) + k\sqrt{1-k^2}]$$

$$= \frac{\pi}{4}\left(\frac{\sin^{-1}(k)}{k} + \sqrt{1-k^2}\right) \tag{4.87}$$

Finally, we give the surface area of the prolate ellipsoid of revolution

$$S = 8abH(0,h)$$

$$= 2\pi ab\left(\frac{\sin^{-1}(k)}{k} + \sqrt{1-k^2}\right) \tag{4.88}$$

$$k = \frac{\sqrt{a^2-b^2}}{a}$$

Particularly, if we assume

$$k \to 0 \tag{4.89}$$

which is reduced to the sphere case of

$$a = b = c = R \tag{4.90}$$

and notice that

$$\lim_{k \to 0} \left[\frac{\sin^{-1} k}{k} \right] = 1 \tag{4.91}$$

$$\lim_{k \to 0} \sqrt{1 - k^2} = 1 \tag{4.92}$$

Equation (4.88) becomes

$$S = S_0 = 4\pi R^2 \tag{4.93}$$

Namely the surface area of the sphere.

Table 4.7 Comparison between complete elliptic and ellipsoidal integrals of the two kinds

	The first kind	The second kind
Complete elliptic integrals	$K(k) = \int_0^1 \dfrac{dx}{\sqrt{1-x^2}\sqrt{1-k^2x^2}}$ $k = \dfrac{\sqrt{a^2-b^2}}{a}$	$E(k) = \int_0^1 \sqrt{\dfrac{1-k^2x^2}{1-x^2}}\,dx$ $k = \dfrac{\sqrt{a^2-b^2}}{a}$
Complete ellipsoidal integrals	$G(k,h) = \int_0^1 \dfrac{dx}{\sqrt{1-h^2x^2}\sqrt{1-k^2x^2}}$ $k = \dfrac{\sqrt{a^2-b^2}}{a}, h = \dfrac{\sqrt{a^2-c^2}}{a}$	$H(e,h) = \int_0^1 \sqrt{1-h^2x^2}\,E\left(\dfrac{e\sqrt{1-x^2}}{\sqrt{1-h^2x^2}}\right)dx$ $e = \dfrac{\sqrt{b^2-c^2}}{b}, h = \dfrac{\sqrt{a^2-c^2}}{a}$

Table 4.8 Comparison between general elliptic and ellipsoidal integrals of the two kinds

	The first kind	The second kind
General elliptic integrals	$u(x,k) = \int_0^x \dfrac{dt}{\sqrt{1-t^2}\sqrt{1-k^2t^2}}$ $k = \dfrac{\sqrt{a^2-b^2}}{a}$	$v(x,k) = \int_0^x \sqrt{\dfrac{1-k^2t^2}{1-t^2}}\,dt$ $k = \dfrac{\sqrt{a^2-b^2}}{a}$
General ellipsoidal integrals	$m(x,k,h) = \int_0^x \dfrac{dt}{\sqrt{1-h^2t^2}\sqrt{1-k^2t^2}}$ $k = \dfrac{\sqrt{a^2-b^2}}{a}, h = \dfrac{\sqrt{a^2-c^2}}{a}$	$n(x,k,h) = \int_0^x \sqrt{1-h^2t^2}\,E\left(\dfrac{e\sqrt{1-t^2}}{\sqrt{1-h^2t^2}}\right)dt$ $e = \dfrac{\sqrt{b^2-c^2}}{b}, h = \dfrac{\sqrt{a^2-c^2}}{a}$

To sum up, we list the comparison between complete elliptic and ellipsoidal integrals of the two kinds in Table 4.7. Table 4.8 pertains to the comparison between general elliptic and ellipsoidal integrals of the two kinds. By comparison, we have

$$m(1,k,h) = G(k,h) \qquad\qquad (4.94)$$

$$n(1,e,h) = H(e,h) \qquad\qquad (4.95)$$

Chapter 5
Theory of Ellipsoidal Function

5.1 Introduction of Ellipsoidal Function

Based on the previous discussion in detail, we have gained a rather clear understanding about the concept of the ellipsoidal integral, especially the ellipsoidal integral of the first kind.

(1) The ellipsoidal integral of the first kind is a simple one-dimensional integral, and it can solve a lot of practical problems.
(2) The capacitance of an arbitrary ellipsoidal conductor can be formulated as:

$$C = \frac{4\pi\varepsilon a}{G(k,h)} \tag{5.1}$$

where

$$G(k,h) = \int_0^1 \frac{dx}{\sqrt{1 - h^2 x^2}\sqrt{1 - k^2 x^2}} \tag{5.2}$$

It is the complete ellipsoidal integral of the first kind.
(3) The ellipsoidal integrals are the reasonable extension of the elliptic integrals, which contain two independent parameters: k and h.

The primary task here is to introduce the ellipsoidal function and enable it to further improve in theory. Considering that we have made a preliminary introduction above, now summarize it again in Table 5.1.

© Science Press, Beijing and Springer Nature Singapore Pte Ltd. 2017
C.-H. Liang, *Notes on the Ellipsoidal Function*,
DOI 10.1007/978-981-10-2908-0_5

Table 5.1 The introduction of ellipsoidal function	General ellipsoidal integral of the first kind $$m = \int_0^x \frac{dx}{\sqrt{1 - h^2 x^2}\sqrt{1 - k^2 x^2}}$$ $$k = \frac{\sqrt{a^2 - b^2}}{a}, h = \frac{\sqrt{a^2 - c^2}}{a}$$	
	Ellipsoidal sine function	$x = s_l(m, k, h)$
	Ellipsoidal cosine h function	$e_l(m, k, h) = \sqrt{1 - h^2 x^2}$
	Ellipsoidal Δ function	$d_l(m, k, h) = \sqrt{1 - k^2 x^2}$

The ellipsoidal functions have two important identical equations:

$$k^2 s_l^2(m, k, h) + d_l^2(m, k, h) = 1 \tag{5.3}$$

$$h^2 s_l^2(m, k, h) + e_l^2(m, k, h) = 1 \tag{5.4}$$

Ellipsoidal cosine function c_l is not defined independently, since s_l, e_l and d_l are complete, and

$$e_l(m, k, 1) = c_l(m, k, 1) \tag{5.5}$$

For simplifying the expressions of the ellipsoidal functions, we omit k and h. Without loss of generality, we choose $a \geq b \geq c$ and the same ellipsoidal parameters of k and h in the following discussion.

5.1.1 The Basic Property of the Ellipsoidal Function

- Special value

$$s_l(0) = 0 \tag{5.6}$$

$$e_l(0) = 1 \tag{5.7}$$

$$d_l(0) = 1 \tag{5.8}$$

- Parity

$$s_l(-u) = -s_l(u) \tag{5.9}$$

$$e_l(-u) = e_l(u) \tag{5.10}$$

$$d_l(-u) = d_l(u) \tag{5.11}$$

In summary, except that s_l is an odd function, e_l and d_l are even functions.

5.1.2   *Derivative of Ellipsoidal Function*

We will discuss the derivatives of the ellipsoidal functions one-by-one here.

(1)
$$\frac{\mathrm{d}}{\mathrm{d}u}(s_l(u)) = d_l(u)e_l(u) \tag{5.12}$$

Proof Obviously, according to $x = s_l(u)$, we have

$$\frac{\mathrm{d}}{\mathrm{d}u}(s_l(u)) = \frac{\mathrm{d}x}{\mathrm{d}u} = \frac{1}{\left(\frac{\mathrm{d}u}{\mathrm{d}x}\right)}$$

Considering

$$\frac{\mathrm{d}u}{\mathrm{d}x} = \frac{1}{\sqrt{1 - h^2x^2}\sqrt{1 - k^2x^2}} = \frac{1}{d_l(u)e_l(u)}$$

the derivative of the ellipsoidal sine function can be obtained.

(2)
$$\frac{\mathrm{d}}{\mathrm{d}u}(e_l(u)) = -h^2 s_l(u)d_l(u) \tag{5.13}$$

Proof Considering the identity

$$h^2 s_l^2(u) + e_l^2(u) = 1$$

and taking the derivative of the above equation with respect to u, we can get

$$2h^2 s_l(u)\frac{\mathrm{d}s_l(u)}{\mathrm{d}u} + 2e_l(u)\frac{\mathrm{d}e_l(u)}{\mathrm{d}u} = 0$$

Substituting Eq. (5.12) into the above equation, we can easily obtain

$$\frac{\mathrm{d}}{\mathrm{d}u}[d_l(u)] = -k^2 s_l(u)e_l(u)$$

(3)
$$\frac{\mathrm{d}}{\mathrm{d}u}(e_l(u)) = -h^2 s_l(u)d_l(u) \tag{5.14}$$

Table 5.2 Comparison of the derivatives of the elliptic and the ellipsoidal functions	Elliptic function	Ellipsoidal function
	$\frac{d}{du}[s_n(u)] = c_n(u)d_n(u)$	$\frac{d}{du}[s_l(u)] = d_l(u)e_l(u)$
	$\frac{d}{du}[c_n(u)] = -s_n(u)d_n(u)$	$\frac{d}{du}[e_l(u)] = -h^2 s_l(u)d_l(u)$
	$\frac{d}{du}[d_n(u)] = -k^2 s_n(u)c_n(u)$	$\frac{d}{du}[d_l(u)] = -k^2 s_l(u)e_l(u)$

Proof Similarly, considering the identity

$$k^2 s_l^2(u) + d_l^2(u) = 1$$

and taking the derivative of the above equation with respect to u, we have

$$2k^2 s_l(u)\frac{ds_l(u)}{du} + 2d_l(u)\frac{dd_l(u)}{du} = 0$$

Inserting Eq. (5.12) into the above equation, we obviously get

$$\frac{d}{du}(d_l(u)) = -k^2 s_l(u)e_l(u)$$

The derivatives of the elliptic functions and the ellipsoidal functions are compared and listed in Table 5.2.

It can be seen from Table 5.2 that the ellipsoidal functions s_l, e_l, and d_l are complete, and their derivatives are self-contained. Hence no ellipsoidal cosine function c_l appears. When comparing derivatives of the ellipsoidal functions with those of the elliptic functions, we could easily learn that the ellipsoidal function e_l corresponds to the elliptic function c_n and, thus, it is reasonable to define $e_l(u)$ as an ellipsoidal cosine h function.

5.1.3   Integral of Ellipsoidal Function

We will discuss the integrals of the ellipsoidal functions one by one.

(1) $$\int s_l(u)du = \pm\frac{1}{kh}\ln[hd_l(u) \mp ke_l(u)] + c \qquad (5.15)$$

Proof Taking derivative of the right hand side of Eq. (5.15), we have

$$\pm\frac{1}{kh}\frac{h\frac{dd_l(u)}{du} \mp k\frac{de_l(u)}{du}}{hd_l(u) \mp ke_l(u)}$$

Considering the following equations

$$\begin{cases} \dfrac{\mathrm{d}d_l(u)}{\mathrm{d}u} = -k^2 s_l(u) e_l(u) \\ \dfrac{\mathrm{d}e_l(u)}{\mathrm{d}u} = -h^2 s_l(u) d_l(u) \end{cases}$$

we can get

$$\pm \frac{1}{kh} \cdot \frac{-k^2 h e_{l(u)} \pm k^2 h d_{l(u)}}{h d_l(u) \mp k e_l(u)} s_l(u)$$

$$= \frac{1}{kh} \cdot \frac{kh(h d_{l(u)} \mp k e_{l(u)})}{h d_l(u) \mp k e_l(u)} s_l(u)$$

$$= s_l(u)$$

(2) $$\int e_l(u)\mathrm{d}u = \frac{i}{k}\ln(d_l(u) - jks_l(u)) + c \tag{5.16}$$

Proof Similarly, taking the derivative of the right-hand side of the above equation, we can get

$$\frac{i}{k}\frac{\dfrac{\mathrm{d}d_l(u)}{\mathrm{d}u} - jk\dfrac{\mathrm{d}s_l(u)}{\mathrm{d}u}}{d_l(u) - jks_l(u)}$$

Substituting

$$\begin{cases} \dfrac{\mathrm{d}d_l(u)}{\mathrm{d}u} = -k^2 s_l(u) e_l(u) \\ \dfrac{\mathrm{d}s_l(u)}{\mathrm{d}u} = e_l(u) d_l(u) \end{cases}$$

into the above equation, we can obtain

$$\frac{d_l(u) - jks_l(u)}{d_l(u) - jks_l(u)} e_l(u) = e_l(u)$$

(3) $$\int d_l(u)\mathrm{d}u = \frac{i}{h}\ln[e_l(u) - jhs_l(u)] + c \tag{5.17}$$

Proof With the similar derivation procedure for the right-hand side of the above equation, we have

Table 5.3 Comparison of integrals of the elliptic functions and the ellipsoidal functions

Elliptic function	Ellipsoidal function
$\int s_n(u)\mathrm{d}u = \pm\frac{1}{k}\ln(d_n(u) \mp kc_n(u)) + C$	$\int s_l(u)\mathrm{d}u = \pm\frac{1}{kh}\ln(hd_l(u) \mp ke_l(u)) + C$
$\int c_n(u)\mathrm{d}u = \frac{j}{k}\ln(d_n(u) - jks_n(u)) + C$	$\int e_l(u)\mathrm{d}u = \frac{j}{k}\ln(d_l(u) - jks_l(u)) + C$
$\int d_n(u)\mathrm{d}u = j\ln(c_n(u) - jks_n(u)) + C$	$\int d_l(u)\mathrm{d}u = \frac{j}{h}\ln(e_l(u) - jhs_l(u)) + C$

$$\frac{i}{h}\frac{\dfrac{\mathrm{d}e_l(u)}{\mathrm{d}u} - jh\dfrac{\mathrm{d}s_l(u)}{\mathrm{d}u}}{e_l(u) - jhs_l(u)}$$

Further considering the following equations

$$\begin{cases} \dfrac{\mathrm{d}e_l(u)}{\mathrm{d}u} = -h^2 s_l(u)d_l(u) \\ \dfrac{\mathrm{d}s_l(u)}{\mathrm{d}u} = e_l(u)d_l(u) \end{cases}$$

we have

$$\frac{i}{h}\frac{-h^2 s_l(u)d_l(u) - jhe_l(u)d_l(u)}{e_l(u) - jhs_l(u)}$$

$$= \frac{e_l(u) - jhs_l(u)}{e_l(u) - jhs_l(u)}d_l(u) = d_l(u)$$

We interestingly compare the integrals of the elliptic functions and ellipsoidal functions in Table 5.3. It can be seen from Table 5.3 that the integrals of the ellipsoidal functions s_l, e_l and d_l are complete, and the ellipsoidal function e_l corresponds to the elliptic function c_n. Moreover, whatever are the elliptic function integrals of c_n and d_n or the ellipsoidal-function integrals of e_l and d_l, they all have generalized spontaneously the real domain to the complex domain, the profound meaning of which is worth discussing.

5.2   Addition Theorem of Ellipsoidal Function

5.2.1   Addition Theorem

Addition theorem is also called addition formula, which will be discussed one by one. We only present the case of "+" here, while the case of "−" can be similarly obtained by considering the parity of s_i, e_i and d_i.

(1)
$$s_l(u+v) = \frac{s_l(u)d_l(v)e_l(v) + s_l(v)d_l(u)e_l(u)}{1 - k^2h^2s_l^2(u)s_l^2(v)}$$
(5.18)

Proof Considering the following definition

$$\begin{cases} u = \int\limits_0^x \frac{dt}{\sqrt{1-h^2t^2}\sqrt{1-k^2t^2}} \\ x = s_l(u,k,h) \\ v = \int\limits_0^y \frac{dt}{\sqrt{1-h^2t^2}\sqrt{1-k^2t^2}} \\ y = s_l(v,k,h) \end{cases}$$

and assuming $u+v = \alpha$ ($\alpha = $ constant), we easily have

$$\left(\frac{dx}{du}\right)^2 = (1-k^2x^2)(1-h^2x^2) = 1 - (k^2+h^2)x^2 + k^2h^2x^4$$
$$\left(\frac{dy}{dv}\right)^2 = \left[\left(\frac{dy}{du}\right)\left(\frac{du}{dv}\right)\right]^2 = \left(\frac{dy}{du}\right)^2$$
$$= 1 - (k^2+h^2)y^2 + k^2h^2y^4$$

In above derivation, the following equation has been considered.

$$\frac{du}{dv} = -1 \text{ or } du+dv = 0$$

Taking the derivatives of both sides of $(dx/du)^2$ and $(dy/du)^2$ with respect to x and y, and dividing the resultant equations by $2(dx/du)$ and $2(dy/du)$, respectively, when (dx/du) and (dy/du) are not equal to zero, we have

$$\begin{cases} \frac{d^2x}{du^2} = -(k^2+h^2)x + 2k^2h^2x^3 \\ \frac{d^2y}{du^2} = -(k^2+h^2)y + 2k^2h^2y^3 \end{cases}$$

Hence, we can get

$$y\frac{d^2x}{du^2} - x\frac{d^2y}{du^2} = \frac{d}{du}\left(y\frac{dx}{du} - x\frac{dy}{du}\right)$$
$$= -2k^2h^2xy(y^2-x^2)$$

and

$$y^2\left(\frac{dx}{du}\right)^2 - x^2\left(\frac{dy}{du}\right)^2 = (y^2-x^2)(1-k^2h^2x^2y^2)$$

Furthermore constructing the following equation

$$\frac{\frac{d}{du}\left[y\frac{dx}{du} - x\frac{dy}{du}\right]}{y\frac{dx}{du} - x\frac{dy}{du}} = -\frac{2k^2h^2xy\left(y\frac{dx}{du} + x\frac{dy}{du}\right)}{1 - k^2h^2x^2y^2}$$

which is equivalent to

$$\frac{d\ln\left[y\frac{dx}{du} - x\frac{dy}{du}\right]}{du} = \frac{d\ln[1 - k^2h^2x^2y^2]}{du}$$

we can obtain

$$\frac{y\frac{dx}{du} - x\frac{dy}{du}}{1 - k^2h^2x^2y^2} = \frac{y(1 - k^2x^2)^{1/2}(1 - h^2x^2)^{1/2} + x(1 - k^2y^2)^{1/2}(1 - h^2y^2)^{1/2}}{1 - k^2h^2x^2y^2}$$

$$= D = \text{constant}$$

Noting that $x = s_l(u)$ and $y = s_l(v)$, we can prove

$$s_l(u + v) = \frac{s_l(u)d_l(v)e_l(v) + s_l(v)d_l(u)e_l(u)}{1 - k^2h^2s_l^2(u)s_l^2(v)}$$

(2) $$e_l(u + v) = \frac{e_l(u)e_l(v) - h^2s_l(u)s_l(v)d_l(u)d_l(v)}{1 - k^2h^2s_l^2(u)s_l^2(v)}$$ (5.19)

Proof For simplicity, we define some simple symbols

$$\begin{cases} s_1 = s_l(u) \\ s_2 = s_l(v) \\ e_1 = e_l(u) \\ e_2 = e_l(v) \\ d_1 = d_l(u) \\ d_2 = d_l(v) \end{cases}$$

In this case, we can have

$$\left(1 - k^2h^2s_1^2s_2^2\right)^2 e_l^2(u + v) = \left(1 - k^2h^2s_1^2s_2^2\right)^2\left[1 - h^2s_l^2(u + v)\right]$$

$$= \left(1 - k^2h^2s_1^2s_2^2\right)^2 - h^2(s_1d_2e_2 + s_2d_1e_1)^2$$

The above equation can be separated into four items as follows:

①
$$\left(1 - k^2 h^2 s_1^2 s_2^2\right)^2 = 1 + k^4 h^4 s_1^4 s_2^4 - 2k^2 h^2 s_1^2 s_2^2$$

②
$$-h^2 s_1^2 d_2^2 e_2^2 = -h^2 s_1^2 \left(1 - k^2 s_2^2\right)\left(1 - h^2 s_2^2\right)$$
$$= \left(-h^2 s_1^2 + k^2 h^2 s_1^2 s_2^2\right)\left(1 - h^2 s_2^2\right)$$
$$= -h^2 s_1^2 + k^2 h^2 s_1^2 s_2^2 - h^4 s_1^2 s_2^2 - k^2 h^4 s_1^2 s_2^4$$

③
$$-h^2 s_2^2 d_1^2 e_1^2 = -h^2 s_2^2 \left(1 - k^2 s_1^2\right)\left(1 - h^2 s_1^2\right)$$
$$= \left(-h^2 s_2^2 + k^2 h^2 s_1^2 s_2^2\right)\left(1 - h^2 s_1^2\right)$$
$$= -h^2 s_2^2 + k^2 h^2 s_1^2 s_2^2 + h^4 s_1^2 s_2^2 - k^2 h^4 s_1^4 s_2^2$$

④
$$- 2h^2 s_1 s_2 d_1 d_2 e_1 e_2$$

Combining ①–④, we can get

$$\boxed{\begin{aligned} &1 - h^2 s_1^2 - h^2 s_2^2 + k^4 h^4 s_1^4 s_2^4 + 2h^4 s_1^2 s_2^2 \\ &- k^2 h^4 s_1^2 s_2^4 - k^2 h^4 s_1^4 s_2^2 - 2h^2 s_1 s_2 d_1 d_2 e_1 e_2 \end{aligned}}$$

On the other side, consider the square of the numerator of the right hand side of Eq. (5.19), i.e.,

$$\left(e_1 e_2 - h^2 s_1 s_2 d_1 d_2\right)^2$$

It can be divided into three terms

⑤
$$e_1^2 e_2^2 = \left(1 - h^2 s_1^2\right)\left(1 - h^2 s_2^2\right)$$
$$= 1 - h^2 s_1^2 - h^2 s_2^2 + h^4 s_1^2 s_2^2$$

⑥
$$h^4 s_1^2 s_2^2 d_1^2 d_2^2 = h^4 s_1^2 s_2^2 \left(1 - k^2 s_1^2\right)\left(1 - k^2 s_2^2\right)$$
$$= \left(h^4 s_1^2 s_2^2 - k^2 h^4 s_1^4 s_2^2\right)\left(1 - k^2 s_2^2\right)$$
$$= h^4 s_1^2 s_2^2 - k^2 h^4 s_1^4 s_2^2 - k^2 h^4 s_1^2 s_2^4 + k^4 h^4 s_1^4 s_2^4$$

⑦
$$- 2s_1 s_2 d_1 d_2 e_1 e_2$$

When combing ⑤–⑦, we have

$$\boxed{\begin{aligned} &1 - h^2 s_1^2 - h^2 s_2^2 + k^4 h^4 s_1^4 s_2^4 + 2h^4 s_1^2 s_2^2 \\ &- k^2 h^4 s_1^2 s_2^4 - k^2 h^4 s_1^4 s_2^2 - 2h^2 s_1 s_2 d_1 d_2 e_1 e_2 \end{aligned}}$$

which is exactly same as the previous one. Hence, we can strictly prove

$$e_l(u+v) = \frac{e_l(u)e_l(v) - h^2 s_l(u)s_l(v)d_l(u)d_l(v)}{1 - k^2 h^2 s_l^2(u)s_l^2(v)}$$

(3) $$d_l(u+v) = \frac{d_l(u)d_l(v) - k^2 s_l(u)s_l(v)e_l(u)e_l(v)}{1 - k^2 h^2 s_l^2(u)s_l^2(v)}$$ (5.20)

Proof A proof similar to that of the equation $e_l(u+v)$ is employed. First, we have

$$\left(1 - k^2 h^2 s_1^2 s_2^2\right)^2 d_l^2(u+v) = \left(1 - k^2 h^2 s_1^2 s_2^2\right)^2 \left[1 - k^2 s_l^2(u+v)\right]$$

Substituting Eq. (5.18) into the above equation, we get

$$\left(1 - k^2 h^2 s_1^2 s_2^2\right)^2 d_l^2(u+v) = \left(1 - k^2 h^2 s_1^2 s_2^2\right)^2 - k^2 (s_1 d_2 e_2 + s_2 d_1 e_1)^2$$

In the exactly same way, we divide the above equation into four items as follows:

① $$\left(1 - k^2 h^2 s_1^2 s_2^2\right)^2 = 1 + k^4 h^4 s_1^4 s_2^4 - 2k^2 h^2 s_1^2 s_2^2$$

②
$$\begin{aligned}
-k^2 s_1^2 d_2^2 e_2^2 &= -k^2 s_1^2 \left(1 - k^2 s_2^2\right)\left(1 - h^2 s_2^2\right) \\
&= \left(-k^2 s_1^2 + k^4 s_1^2 s_2^2\right)\left(1 - h^2 s_2^2\right) \\
&= -k^2 s_1^2 + k^2 h^2 s_1^2 s_2^2 + k^4 s_1^2 s_2^2 - k^4 h^2 s_1^2 s_2^4
\end{aligned}$$

③
$$\begin{aligned}
-k^2 s_2^2 d_1^2 e_1^2 &= -k^2 s_2^2 \left(1 - k^2 s_1^2\right)\left(1 - h^2 s_1^2\right) \\
&= \left(-k^2 s_2^2 + k^4 s_1^2 s_2^2\right)\left(1 - h^2 s_1^2\right) \\
&= -k^2 s_2^2 + k^2 h^2 s_1^2 s_2^2 + k^4 s_1^2 s_2^2 - k^4 h^2 s_1^4 s_2^2
\end{aligned}$$

④ $$- 2k^2 s_1 s_2 d_1 d_2 e_1 e_2$$

With combination of ①–④, we can get

$$\boxed{\begin{aligned}
&1 + k^4 h^4 s_1^4 s_2^4 - k^2 s_1^2 - k^2 s_2^2 + 2k^4 s_1^2 s_2^2 \\
&- k^4 h^2 s_1^4 s_2^2 - k^4 h^2 s_1^2 s_2^4 - 2k^2 s_1 s_2 d_1 d_2 e_1 e_2
\end{aligned}}$$

On the other hand, the square of the numerator of Eq. (5.20) can be expressed as

$$\left(d_1 d_2 - k^2 s_1 s_2 e_1 e_2\right)^2$$

Table 5.4 Comparison of addition formulas between the elliptic functions and the ellipsoidal functions

Elliptic function	Ellipsoidal function
$s_n(u+v) = \frac{s_n(u)c_n(v)d_n(v) + s_n(v)c_n(u)d_n(u)}{1 - k^2 s_n^2(u) s_n^2(v)}$	$s_l(u+v) = \frac{s_l(u)d_l(v)e_l(v) + s_l(v)d_l(u)e_l(u)}{1 - k^2 s_l^2(u) s_l^2(v)}$
$c_n(u+v) = \frac{c_n(u)c_n(v) - s_n(u)s_n(v)d_n(u)d_n(v)}{1 - k^2 s_n^2(u) s_n^2(v)}$	$e_l(u+v) = \frac{e_l(u)e_l(v) - h^2 s_l(u)s_l(v)d_l(u)d_l(v)}{1 - k^2 h^2 s_l^2(u) s_l^2(v)}$
$d_n(u+v) = \frac{d_n(u)d_n(v) - k^2 s_n(u)s_n(v)c_n(u)c_n(v)}{1 - k^2 s_n^2(u) s_n^2(v)}$	$d_l(u+v) = \frac{d_l(u)d_l(v) - k^2 s_l(u)s_l(v)e_l(u)e_l(v)}{1 - k^2 h^2 s_l^2(u) s_l^2(v)}$

Similarly, we separate it into three items as

⑤
$$d_1^2 d_2^2 = \left(1 - k^2 s_1^2\right)\left(1 - k^2 s_2^2\right)$$
$$= 1 - k^2 s_1^2 - k^2 s_2^2 + k^4 s_1^2 s_2^2$$

⑥
$$k^4 s_1^2 s_2^2 e_1^2 e_2^2 = k^4 s_1^2 s_2^2 \left(1 - h^2 s_1^2\right)\left(1 - h^2 s_2^2\right)$$
$$= \left(k^4 s_1^2 s_2^2 - k^4 h^2 s_1^4 s_2^2\right)\left(1 - h^2 s_2^2\right)$$
$$= k^4 s_1^2 s_2^2 - k^4 h^2 s_1^4 s_2^2 - k^4 h^2 s_1^2 s_2^4 + k^4 h^4 s_1^4 s_2^4$$

⑦
$$- 2k^2 s_1 s_2 d_1 d_2 e_1 e_2$$

With combination of ⑤–⑦, we can obtain

$$\boxed{\begin{aligned}1 + k^4 h^4 s_1^4 s_2^4 - k^2 s_1^2 - k^2 s_2^2 + 2k^4 s_1^2 s_2^2 \\ - k^4 h^2 s_1^4 s_2^2 - k^4 h^2 s_1^2 s_2^4 - 2k^2 s_1 s_2 d_1 d_2 e_1 e_2\end{aligned}}$$

By comparison, we can know that they are completely same. Hence, we can prove

$$d_l(u+v) = \frac{d_l(u)d_l(v) - k^2 s_l(u)s_l(v)e_l(u)e_l(v)}{1 - k^2 h^2 s_l^2(u) s_l^2(v)}$$

Similarly, we compare the addition formulas of the elliptic functions and the ellipsoidal functions in Table 5.4. From Table 5.4, we learned again that s_l, e_l, and d_l have already formed a complete system of the ellipsoidal functions, and the ellipsoidal function e_l is in corresponding to the elliptic function c_n. Hence, it is unnecessary to define the ellipsoidal cosine function.

5.2.2 Double Argument Formula

As the most common cases, we further extend the addition formulas of the ellipsoidal functions to the double argument formulas. In other words, we have

$$v = u \ \text{ or } \ u + v = 2u \tag{5.21}$$

(1) Ellipsoidal sine function

$$s_l(2u) = \frac{2s_l(u)d_l(u)e_l(u)}{1 - k^2h^2s_l^4(u)} \tag{5.22}$$

Substituting Eq. (5.21) into Eq. (5.18), we can obtain Eq. (5.22).

(2) Ellipsoidal cosine h function

$$e_l(2u) = 1 - \frac{2h^2s_l^2(u)d_l^2(u)}{1 - k^2h^2s_l^4(u)} \tag{5.23}$$

Proof Inserting $u + v = 2u$ into Eq. (5.19), we have

$$e_l(2u) = \frac{e^2 - h^2s^2d^2}{1 - k^2h^2s^4}$$

The numerator of the above equation is rewritten as

$$\begin{aligned}
\left(1 - h^2s^2\right) - h^2s^2\left(1 - k^2s^2\right) \\
= 1 - 2h^2s^2 + k^2h^2s^4 \\
= \left(1 - k^2h^2s^4\right) - 2h^2s^2 + 2k^2h^2s^4 \\
= \left(1 - k^2h^2s^4\right) - 2h^2s^2\left(1 - k^2s^2\right) \\
= \left(1 - k^2h^2s^4\right) - 2h^2s^2d^2
\end{aligned}$$

Further considering the denominator, we can strictly prove Eq. (5.23) as

$$e_l(2u) = 1 - \frac{2h^2s_l^2(u)d_l^2(u)}{1 - k^2h^2s_l^4(u)}$$

(3) Ellipsoid Δ function

$$d_l(2u) = 1 - \frac{2k^2s_l^2(u)e_l^2(u)}{1 - k^2h^2s_l^4(u)} \tag{5.24}$$

Table 5.5 Comparison of double argument formulas between the elliptic functions and the ellipsoidal functions

Elliptic function	Ellipsoidal function
$s_n(2u) = \frac{2s_n(u)c_n(u)d_n(u)}{1-k^2s_n^4(u)}$	$s_l(2u) = \frac{2s_l(u)d_l(u)e_l(u)}{1-k^2h^2s_l^4(u)}$
$c_n(2u) = 1 - \frac{2s_n^2(u)d_n^2(u)}{1-k^2s_n^4(u)}$	$e_l(2u) = 1 - \frac{2h^2s_l^2(u)d_l^2(u)}{1-k^2h^2s_l^4(u)}$
$d_n(2u) = 1 - \frac{2k^2s_n^2(u)c_n^2(u)}{1-k^2s_n^4(u)}$	$d_l(2u) = 1 - \frac{2k^2s_l^2(u)e_l^2(u)}{1-k^2h^2s_l^4(u)}$

Proof Substituting $v = u$ into Eq. (5.20), we can get

$$d_l(2u) = \frac{d^2 - k^2s^2e^2}{1 - k^2h^2s^4}$$

We rewrite the numerator of the above equation as

$$\left(1 - k^2s^2\right) - k^2s^2\left(1 - h^2s^2\right) = \left(1 - k^2h^2s^4\right) - 2k^2s^2e^2$$

Combining with the denominator, we can strictly prove Eq. (5.24). Table 5.5 has listed the comparison of the double argument formula between the elliptic functions and the ellipsoidal functions.

5.3   Periodicity of the Ellipsoidal Function

An important feature of the ellipsoidal functions is their periodicities. We will discuss it in two aspects: the real periods of the ellipsoidal functions and their imaginary periods. Assuming $a \geq b \geq c$ as before, and

$$k = \frac{\sqrt{a^2 - b^2}}{a}, \quad h = \frac{\sqrt{a^2 - c^2}}{a}$$

we have

$$h \geq k \tag{5.25}$$

5.3.1   The Real Period of the Ellipsoidal Function

We define

$$m = \int_0^{\frac{1}{h}} \frac{\mathrm{d}t}{\sqrt{1 - h^2t^2}\sqrt{1 - k^2t^2}} \tag{5.26}$$

Making the following substitution

$$
\begin{cases}
t' = \frac{t}{h} \\
\mathrm{d}t' = \frac{\mathrm{d}t}{h} \\
t \in \left[0, \frac{1}{h}\right] \text{ corresponding to } t_1 \in [0, 1]
\end{cases}
\tag{5.27}
$$

we can get

$$
m = \int_0^1 \frac{\mathrm{d}t'}{\sqrt{1 - t'^2}\sqrt{1 - \left(\frac{k}{h}\right)^2 t'^2}}
\tag{5.28}
$$

According to the definition of the complete elliptic integral of the first kind, e.g.,

$$
K(g) = \int_0^1 \frac{\mathrm{d}x}{\sqrt{1 - x^2}\sqrt{1 - g^2 x^2}}
\tag{5.29}
$$

we can know

$$
m = \frac{1}{h} K(g), g = \left(\frac{k}{h}\right)
\tag{5.30}
$$

Considering some special values

$$
\begin{cases}
s_l(m, k, h) = \frac{1}{h} \\
e_l(m, k, h) = 0 \\
d_l(m, k, h) = \sqrt{1 - g^2} = g'
\end{cases}
\tag{5.31}
$$

we can obtain from the above addition theorem

$$
\begin{aligned}
s_l(u + m) &= \frac{s_l(u) d_l(m) e_l(m) + s_l(m) d_l(u) e_l(u)}{1 - k^2 h^2 s_l^2(u) s_l^2(m)} \\
&= \frac{\frac{1}{h} d_l(u) e_l(u)}{1 - k^2 s_l^2(u)} \\
&= \frac{1}{h} \frac{e_l(u)}{d_l(u)}
\end{aligned}
\tag{5.32}
$$

$$\begin{aligned}
e_l(u+m) &= \frac{e_l(u)e_l(m) - h^2 s_l(u)s_l(m)d_l(u)d_l(m)}{1 - k^2 h^2 s_l^2(u)s_l^2(m)} \\
&= -\frac{hg's_l(u)d_l(u)}{1 - d_l^2(u)} \\
&= -\frac{hg's_l(u)}{d_l(u)}
\end{aligned} \tag{5.33}$$

$$\begin{aligned}
d_l(u+m) &= \frac{d_l(u)d_l(m) - k^2 s_l(u)s_l(m)e_l(u)e_l(m)}{1 - k^2 h^2 s_l^2(u)s_l^2(m)} \\
&= \frac{g'd_l(u)}{d_l^2(u)} \\
&= \frac{g'}{d_l(u)}
\end{aligned} \tag{5.34}$$

Hence, the formula of $(u+m)$ can be summarized as

$$e_l(u+m) = -\frac{hg's_l(u)}{d_l(u)}$$

$$s_l(u+m) = \frac{1}{h}\frac{e_l(u)}{d_l(u)} \tag{5.35}$$

$$d_l(u+m) = \frac{g'}{d_l(u)}$$

To study the real periods of the ellipsoidal functions, we further present

$$\begin{aligned}
s_l(u+2m) &= s_l[(u+m)+m] \\
&= \frac{1}{h}\frac{e_l(u+m)}{d_l(u+m)} \\
&= \frac{1}{h}\frac{-hg'\frac{s_l(u)}{d_l(u)}}{g'\frac{1}{d_l(u)}} = -s_l(u)
\end{aligned} \tag{5.36}$$

$$\begin{aligned}
e_l(u+2m) &= e_l[(u+m)+m] \\
&= -hg'\frac{s_l(u+m)}{d_l(u+m)} \\
&= -hg'\frac{\frac{e_l(u)}{hd_l(u)}}{g'\frac{1}{d_l(u)}} = -e_l(u)
\end{aligned} \tag{5.37}$$

Table 5.6 The real period of the ellipsoidal function

u	$(u+m)$	$(u+2m)$	$(u+4m)$
$s_l(u)$	$s_l(u+m) = \frac{1}{h}\frac{e_l(u)}{d_l(u)}$	$s_l(u+2m) = -s_l(u)$	$s_l(u+4m) = s_l(u)$
$e_l(u)$	$e_l(u+m) = -\frac{hg's_l(u)}{d_l(u)}$	$e_l(u+2m) = -e_l(u)$	$e_l(u+4m) = e_l(u)$
$d_l(u)$	$d_l(u+m) = \frac{g'}{d_l(u)}$	$d_l(u+2m) = d_l(u)$	

$$d_l(u+2m) = d_l[(u+m)+m]$$

$$= \frac{g'}{d_l(u+m)}$$

$$= \frac{g'}{g'\frac{1}{d_l(u)}} = d_l(u) \tag{5.38}$$

Finally, we can obtain

$$s_l(u+4m) = -s_l(u+2m) = s_l(u)$$

$$e_l(u+4m) = -e_l(u+2m) = e_l(u) \tag{5.39}$$

$$d_l(u+2m) = d_l(u)$$

For the elliptical functions, the real periods of $s_l(u)$ and $e_l(u)$ are $4m$, while the real period of $d_l(u)$ is $2m$, as shown in Table 5.6.

5.3.2   The Imaginary Period of the Ellipsoidal Function

Using the imaginary number, we can express the ellipsoidal functions as follows:

$$jn = \int_0^{jy} \frac{dt}{\sqrt{1-h^2t^2}\sqrt{1-k^2t^2}} \tag{5.40}$$

which is equivalent to

$$jy = s_l(jn,k,h) \tag{5.41}$$

where n and y are real numbers.

With the following substitution

$$t = j\frac{t_1}{\sqrt{1 - h^2 t_1^2}} \tag{5.42}$$

$$dt = j\frac{h^2 t_1^2 + (1 - h^2 t_1^2)}{\left(1 - h^2 t_1^2\right)^{3/2}} dt_1$$
$$= j\frac{dt_1}{\sqrt{1 - h^2 t_1^2}} \tag{5.43}$$

and

$$y = \frac{s}{\sqrt{1 - h^2 s^2}} \quad 0 \le s < 1 \tag{5.44}$$

we have

$$\sqrt{1 - h^2 t^2} = \sqrt{1 + \frac{h^2 t_1^2}{1 - h^2 t_1^2}} = \frac{1}{\sqrt{1 - h^2 t_1^2}} \tag{5.45}$$

and

$$\sqrt{1 - k^2 t^2} = \sqrt{1 + \frac{k^2 t_1^2}{1 - h^2 t_1^2}}$$
$$= \frac{\sqrt{1 - (h^2 - k^2) t_1^2}}{\sqrt{1 - h^2 t_1^2}}$$
$$= \frac{\sqrt{1 - \tilde{k}^2 t_1^2}}{\sqrt{1 - h^2 t_1^2}} \tag{5.46}$$

where

$$\tilde{k} = \sqrt{h^2 - k^2} \tag{5.47}$$

Therefore, the integrand in Eq. (5.30) can be specifically expressed as

$$\frac{dt}{\sqrt{1-h^2t^2}\sqrt{1-k^2t^2}} = \frac{\dfrac{jdt_1}{\left(1-h^2t_1^2\right)^{3/2}}}{\dfrac{1}{\left(1-h^2t_1^2\right)^{1/2}}\dfrac{\left(1-\tilde{k}^2t_1^2\right)^{1/2}}{\left(1-h^2t_1^2\right)^{1/2}}}$$

$$= j\frac{dt_1}{\sqrt{\left(1-h^2t_1^2\right)}\sqrt{\left(1-\tilde{k}^2t_1^2\right)}}$$

(5.48)

Considering that an integral is irrelevant to the expressions of the integral variables, we have

$$n = \int_0^s \frac{dt}{\sqrt{1-h^2t^2}\sqrt{1-\tilde{k}^2t^2}}$$

(5.49)

which is equivalent to

$$s = s_l\left(n,\tilde{k},h\right)$$

(5.50)

We rewrite Eq. (5.41) as

$$s_l(jn,k,h) = j\frac{s_l\left(n,\tilde{k},h\right)}{e_l\left(n,\tilde{k},h\right)}$$

(5.51)

Considering

$$e_l^2(jn,k,h) = 1 - h^2 s_l(jn,k,h)$$

$$= 1 + \frac{h^2 s_l^2\left(n,\tilde{k},h\right)}{e_l^2\left(n,\tilde{k},h\right)}$$

$$= \frac{1 - h^2 s_l^2\left(n,\tilde{k},h\right) + h^2 s_l^2\left(n,\tilde{k},h\right)}{e_l^2\left(n,\tilde{k},h\right)}$$

$$= \frac{1}{e_l^2\left(n,\tilde{k},h\right)}$$

we can get

$$e_l(jn,k,h) = \frac{1}{e_l\left(n,\tilde{k},h\right)}$$

(5.52)

Table 5.7 Elliptic function and ellipsoidal function based on imaginary arguments

	Sine	Cosine	Δ
Elliptic function $k' = \sqrt{1-k^2}$	$s_n(jv,k) = j\frac{s_n(v,k')}{c_n(v,k')}$	$c_n(jv,k) = \frac{1}{c_n(v,k')}$	$d_n(jv,k) = \frac{d_n(v,k')}{c_n(v,k')}$
Ellipsoidal function $\tilde{k} = \sqrt{h^2-k^2}$	$s_l(jn,k,h) = j\frac{s_l(n,\tilde{k},h)}{e_l(n,\tilde{k},h)}$	$e_l(jn,k,h) = \frac{1}{e_l(n,\tilde{k},h)}$	$d_l(jn,k,h) = \frac{d_l(n,\tilde{k},h)}{e_l(n,\tilde{k},h)}$

Further we consider

$$
d_l^2(jn,k,h) = 1 - k^2 s_l^2(jn,k,h)
$$
$$
= 1 + \frac{k^2 s_l^2(n,\tilde{k},h)}{e_l^2(n,\tilde{k},h)} \tag{5.53}
$$

which can be reduced to

$$
d_l^2(jn,k,h) = \frac{1 - h^2 s_l^2(n,\tilde{k},h) + k^2 s_l^2(n,\tilde{k},h)}{e_l^2(n,\tilde{k},h)}
$$
$$
= \frac{1 - \tilde{k}^2 s_l^2(n,\tilde{k},h)}{e_l^2(n,\tilde{k},h)} = \frac{d_l^2(n,\tilde{k},h)}{e_l^2(n,\tilde{k},h)}
$$

Hence we have

$$
d_l(jn,k,h) = \frac{d_l(n,\tilde{k},h)}{e_l(n,\tilde{k},h)} \tag{5.54}
$$

The summary is given in Table 5.7.

It can be proved that the imaginary period of $s_l(jn,k,h)$ is $j2m'$, while the imaginary period of $d_l(jn,k,h)$ and $e_l(jn,k,h)$ is $j4m'$, where

$$
\tilde{m}' = \int_0^{\frac{1}{h}} \frac{dt}{\sqrt{1 - h^2 t^2}\sqrt{1 - \tilde{k}^2 t^2}} \tag{5.55}
$$

Table 5.8 Complex argument expression of elliptic function $k' = \sqrt{1-k^2}$

$$
s_n(u+jv,k) = \frac{s_n(u,k)d_n(v,k') + jc_n(u,k)d_n(u,k)s_n(v,k')c_n(v,k')}{c_n^2(v,k') + k^2 s_n^2(u,k)s_n^2(v,k')}
$$

$$
c_n(u+jv,k) = \frac{c_n(u,k)c_n(v,k') - js_n(u,k)d_n(u,k)s_n(v,k')d_n(v,k')}{c_n^2(v,k') + k^2 s_n^2(u,k)s_n^2(v,k')}
$$

$$
d_n(u+jv,k) = \frac{d_n(u,k)c_n(v,k')d_n(v,k') - jk^2 s_n(u,k)c_n(u,k)s_n(v,k')}{c_n^2(v,k') + k^2 s_n^2(u,k)s_n^2(v,k')}
$$

Table 5.9 Complex argument expression of ellipsoidal function $\tilde{k} = \sqrt{h^2 - k^2}$

$$s_l(m+jn,k,h) = \frac{s_l(m,k,h)d_l\left(m,\tilde{k},h\right) + je_l(m,k,h)d_l(m,k,h)s_l\left(n,\tilde{k},h\right)e_l\left(n,\tilde{k},h\right)}{e_l^2\left(n,\tilde{k},h\right) + k^2h^2s_l^2(m,k,h)s_l^2\left(n,\tilde{k},h\right)}$$

$$d_l(m+jn,k,h) = \frac{d_l(m,k,h)d_l\left(m,\tilde{k},h\right) - jk^2s_l(m,k,h)e_l(m,k,h)s_l\left(n,\tilde{k},h\right)}{e_l^2\left(n,\tilde{k},h\right) + k^2h^2s_l^2(m,k,h)s_l^2\left(n,\tilde{k},h\right)}$$

$$e_l(m+jn,k,h) = \frac{e_l(m,k,h)e_l\left(n,\tilde{k},h\right) - js_l(m,k,h)d_l\left(m,\tilde{k},h\right)s_l\left(n,\tilde{k},h\right)d_l\left(n,\tilde{k},h\right)}{e_l^2\left(n,\tilde{k},h\right) + k^2h^2s_l^2(m,k,h)s_l^2\left(n,\tilde{k},h\right)}$$

Applying the imaginary arguments into the addition theorem, we can obtain the corresponding expressions based on complex arguments. Tables 5.8 and 5.9 list the complex argument expressions of the elliptic functions and the ellipsoidal functions, respectively.

Chapter 6
Conformal Mapping of Ellipsoidal Integral

6.1 Conformal Mapping of Ellipsoidal Integral

Consider complex upper half-plane $z = x + jy$. With the following relationship

$$z = s_l(w, k, h) \tag{6.1}$$

conformal mapping of ellipsoid functions can be defined in the corresponding complex plane $w = u + jv$ as

$$w = \int_0^z \frac{\mathrm{d}t}{\sqrt{1 - h^2 t^2}\sqrt{1 - k^2 t^2}} \tag{6.2}$$

which is rewritten as

$$
\begin{aligned}
\frac{\mathrm{d}w}{\mathrm{d}z} &= \frac{1}{(1 - h^2 z^2)^{1/2}(1 - k^2 z^2)^{1/2}} \\
&= \frac{1}{kh}\left(z + \frac{1}{k}\right)^{-1/2}\left(z - \frac{1}{k}\right)^{-1/2}\left(z + \frac{1}{h}\right)^{-1/2}\left(z - \frac{1}{h}\right)^{-1/2}
\end{aligned}
\tag{6.3}
$$

Assume $a \geq b \geq c$ and

$$k = \frac{\sqrt{a^2 - b^2}}{a}, \quad h = \frac{\sqrt{a^2 - c^2}}{a}$$

we have

$$h \geq k \quad \text{or} \quad \frac{1}{k} \geq \frac{1}{h} \tag{6.4}$$

© Science Press, Beijing and Springer Nature Singapore Pte Ltd. 2017
C.-H. Liang, *Notes on the Ellipsoidal Function*,
DOI 10.1007/978-981-10-2908-0_6

The above Eq. (6.3) is well-known Schwarz–Christoffel mapping. And a general expression can be given by

$$\frac{dw}{dz} = A \prod_i (z - a_i)^{\frac{\alpha_i}{\pi} - 1} \tag{6.5}$$

The upper half-plane of complex independent variant z is transformed into the interior of a polygon of complex dependent w. Here α_i in Eq. (6.5) denotes interior angles of the polygon in the w plane. Specifically α_i in Eq. (6.3) becomes

$$\alpha_i = \frac{\pi}{2} \quad (i = 1, 2, 3, 4) \tag{6.6}$$

In other words, the ellipsoid functions map the upper half-plane of the z plane to a rectangular of the w plane.

Case 1 $z = \frac{1}{h}$

$$w = \int_0^{\frac{1}{h}} \frac{dt}{\sqrt{1 - h^2 t^2}\sqrt{1 - k^2 t^2}} = M \tag{6.7}$$

This is because we assume

$$M = \frac{1}{h} \int_0^1 \frac{dt_1}{\sqrt{1 - t_1^2}\sqrt{1 - g^2 t_1^2}} \tag{6.8}$$

where

$$g = \left(\frac{k}{h}\right) \leq 1 \tag{6.9}$$

Further we can obtain

$$M = \frac{1}{h} K(g) \tag{6.10}$$

In Eq. (6.10), K represents the complete elliptic integral.

Case 2 $z = -\frac{1}{h}$

According to the parity property of the ellipsoidal sine function s_l, i.e.,

$$-z = -s_l(w) = s_l(-w) \tag{6.11}$$

we have

$$w = -M \tag{6.12}$$

Case 3 $z = \frac{1}{k}$

Again we write

$$w = \int_0^{\frac{1}{k}} \frac{dt}{\sqrt{1 - h^2 t^2}\sqrt{1 - k^2 t^2}} \tag{6.13}$$

which can be reexpressed as

$$w = \int_0^{\frac{1}{h}} \frac{dt}{\sqrt{1 - h^2 t^2}\sqrt{1 - k^2 t^2}} + \int_{\frac{1}{h}}^{\frac{1}{k}} \frac{dt}{\sqrt{1 - h^2 t^2}\sqrt{1 - k^2 t^2}} = M + N \tag{6.14}$$

We emphasize on the derivation of the integral N. Introducing a variable t_1 according to the following relationship

$$\begin{cases} t_1 = 0 & \leftrightarrow & t = \frac{1}{h} \\ t_1 = 1 & \leftrightarrow & t = \frac{1}{k} \end{cases} \tag{6.15}$$

and considering

$$g' = \sqrt{1 - g^2} \tag{6.16}$$

we have

$$t = \frac{1}{h\sqrt{1 - g'^2 t_1^2}} \tag{6.17}$$

Note that Eq. (6.17) satisfies the conditions given in Eq. (6.15).
According to Eq. (6.17), we can obtain

$$dt = \frac{g'^2 t_1 dt_1}{h\left(1 - g'^2 t_1^2\right)^{3/2}} \tag{6.18}$$

Considering

$$\left(1 - h^2 t^2\right) = -\left(h^2 t^2 - 1\right)$$

$$= -\left[\frac{h^2}{h^2\left(1 - g'^2 t_1^2\right)} - 1\right]$$

$$= -\frac{g'^2 t_1^2}{\left(1 - g'^2 t_1^2\right)} \tag{6.19}$$

we can get

$$\sqrt{1 - h^2 t^2} = -j\frac{g' t_1}{\sqrt{1 - g'^2 t_1^2}} \tag{6.20}$$

Similarly, considering

$$\left(1 - k^2 t^2\right) = 1 - \frac{k^2}{h^2\left(1 - g'^2 t_1^2\right)} = 1 - \frac{g^2}{1 - g'^2 t_1^2}$$

$$= \frac{1 - g'^2 t_1^2 - g^2}{1 - g'^2 t_1^2} = \frac{g'^2\left(1 - t_1^2\right)}{1 - g'^2 t_1^2} \tag{6.21}$$

we have

$$\sqrt{1 - k^2 t^2} = g'\frac{\sqrt{1 - t_1^2}}{\sqrt{1 - g'^2 t_1^2}} \tag{6.22}$$

Substituting Eqs. (6.18), (6.20), and (6.22) into Eq. (6.14), we obtain

$$N = \int_{\frac{1}{h}}^{\frac{1}{k}} \frac{dt}{\sqrt{1 - h^2 t^2}\sqrt{1 - k^2 t^2}}$$

$$= j\int_0^1 \frac{\frac{g'^2 t_1}{h\left(1 - g'^2 t_1^2\right)^{3/2}}\, dt_1}{\frac{g'\sqrt{1 - t_1^2}}{\sqrt{1 - g'^2 t_1^2}} \cdot \frac{g' t_1}{\sqrt{1 - g'^2 t_1^2}}}$$

$$= \frac{j}{h}\int_0^1 \frac{dt_1}{\sqrt{1 - t_1^2}\sqrt{1 - g'^2 t_1^2}}$$

$$= \frac{j}{h} K'(g') \tag{6.23}$$

where

$$K'(g') = \int\limits_0^1 \frac{\mathrm{d}t}{\sqrt{1-t^2}\sqrt{1-g'^2 t^2}} \tag{6.24}$$

Further assuming

$$M' = \frac{K'(g')}{h} \tag{6.25}$$

the integral w can be obtained as follows:

$$
\begin{aligned}
w &= \int\limits_0^{\frac{1}{k}} \frac{\mathrm{d}t}{\sqrt{1-h^2 t^2}\sqrt{1-k^2 t^2}} \\
&= M + N = M + jM' \\
&= \frac{1}{h}\left(K(g) + jK'(g')\right)
\end{aligned}
\tag{6.26}
$$

Very clearly, real and imaginary parts of w are tightly relevant to the elliptic integral.

Case 4 $z = -\frac{1}{k}$

According to the odd symmetry, we easily have

$$
\begin{aligned}
w &= -M + N = -M + jM' \\
&= \frac{1}{h}\left(-K(g) + jK'(g')\right)
\end{aligned}
\tag{6.27}
$$

Rectangular conformal mapping from the z upper half-plane to the w plane is shown in Fig. 6.1.

5.2   Dual Periodicity of Ellipsoidal Function

First, we give the definition of dual periodicity of ellipsoidal function.

Definition If for an arbitrary complex variable z, there is a complex function $f(z)$ which satisfies

$$f(z + 2\omega_1) = f(z) \tag{6.28}$$

Fig. 6.1 Ellipsoidal
conformal mapping
$z = s_l(w, k, h)$

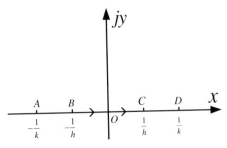

(a) $z = x + jy$ upper half plane

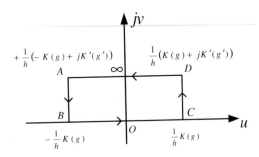

(b) $w = u + jv$ rectangular region

$$f(z + 2\omega_2) = f(z) \tag{6.29}$$

in which the ratio of two complex numbers ω_2/ω_1 is not pure imaginary, the function $f(z)$ is defined as a dual periodic function.

Now, in a parallelogram with four vertices O, $2\omega_1$, $2\omega_2$, and $2\omega_1 + 2\omega_2$, if there are not two complex numbers Q_1 and Q_2 satisfying the following the condition

$$f(z + Q_1) = f(z + Q_2) \tag{6.30}$$

the complex numbers ω_1 and ω_2 are defined as the dual periodicity, and the corresponding parallelogram is called fundamental parallelogram, as shown in Fig. 6.2.

We concentrate on the ellipsoidal sine function s_l, which is a dual periodic function with the dual periodicity of $4M$ and $j2M'$, as shown in Fig. 6.3.

We notice that the ellipsoidal sine function has

$$z = s_l(\omega) \tag{6.31}$$

$$\omega = u + jv \tag{6.32}$$

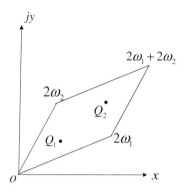

Fig. 6.2 Dual periodic function in complex function and fundamental parallelogram

$$z = \frac{1}{h} \quad \omega = M = \frac{1}{h}K(g)$$

$$z = \frac{1}{k} \quad \omega = M + jM' = \frac{1}{h}(K(g) + jK'(g')) \tag{6.33}$$

in which $g = k/h$ and $g' = \sqrt{1 - g^2}$. We have known that the complete elliptical functions $K(g)$ and $K'(g')$ are periodic functions, which means that the complete elliptical functions have dual periodicity. Similarly, the conclusion can be extended to the ellipsoidal functions, e.g.,

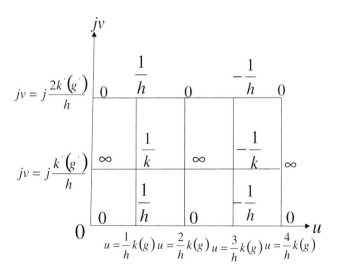

Fig. 6.3 Dual periodicity and fundamental parallelogram of the ellipsoidal sine function $z = s_l(\omega)$, $\omega = u + jv$

$$\omega = M + jM' \qquad\qquad (6.34)$$

where M and M' must be periodic functions. Further, the ellipsoidal functions have dual periodicity, as shown in Fig. 6.3.

Chapter 7
Network of Ellipsoidal Function

Network of ellipsoidal function is an important application of the ellipsoidal function in the field of electromagnetics. The function of the network is to use the ellipsoidal function for synthesis of the tri-band filter with superior performance.

7.1 Design of Filter Based on Ellipsoidal Function

The main procedure for the design of the filter is: The desired model is first determined, and then approximate function is used to realize the performance of the model. For an ellipsoidal-function filter, a tri-band model is proposed here. As shown in Fig. 7.1, passband is $\omega \in [0, \omega_{s1}]$, transition band is $\omega \in [\omega_{s1}, \omega_{s2}]$, stopband is $\omega \in [\omega_{s2}, \infty]$, and $G(\omega^2)$ denotes the gain response of the filter.

We write the gain response function in a general form as

$$G(\omega^2) = \frac{H_n}{1 + \varepsilon^2 F_n^2(\omega)} \tag{7.1}$$

In the denominator of the above fraction, the constraint condition of $F_n(\omega)$ is

$$F_n(\omega) = \frac{U_n(\omega)}{V_n(\omega)} \tag{7.2}$$

where both $U_n(\omega)$ and $V_n(\omega)$ are integral-order polynomial functions of the frequency, and

$$G(0) = H_n \tag{7.3}$$

$$G(\omega_{s1}) = \frac{H_n}{1 + \varepsilon^2} \tag{7.4}$$

© Science Press, Beijing and Springer Nature Singapore Pte Ltd. 2017
C.-H. Liang, *Notes on the Ellipsoidal Function*,
DOI 10.1007/978-981-10-2908-0_7

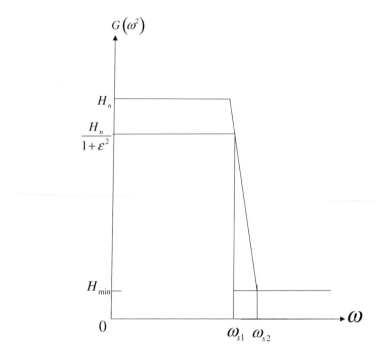

Fig. 7.1 Perfect low pass model of the filter with a tri-band response

This means that

$$F_n(0) = 0 \tag{7.5}$$

$$F_n(\omega_{s1}) = 1 \tag{7.6}$$

Figure 7.1 shows a low pass model. When $\omega > \omega_{s2}$, we have

$$G(\omega^2) \leq H_{\min} \tag{7.7}$$

Figure 7.2 demonstrates the synthesis procedure of the ellipsoidal-function filter

7.2   Analysis of Ellipsoidal-Function Filter

When studying the conformal mapping of the ellipsoidal function, we carefully notice that the conformal mapping of $z = s_l(\omega)$ implicitly corresponds to the triple bands, i.e., passband, transition band, and stopband, as shown in Fig. 7.3.

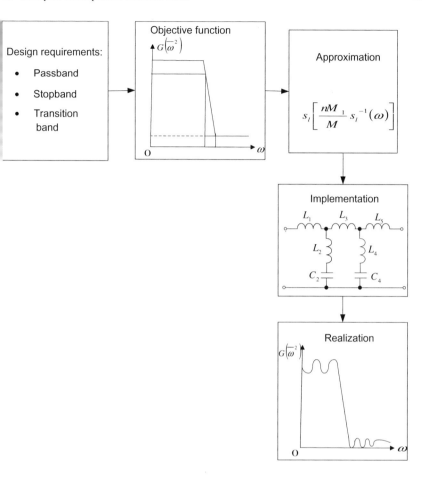

Fig. 7.2   Design procedure of the ellipsoidal-function filter

The approximation function is essentially designed in a comparison way. We compare the approximation function with perfect model. Table 7.1 shows a comparison between Chebyshev synthesis and Ellipsoidal synthesis.

In practice, the approximation function based on the elliptical function in current engineering has been proposed, i.e.,

$$s_n \left[\frac{nk_1}{k} s_n^{-1}(\overline{\omega}), k_1 \right] \tag{7.8}$$

The question about the ellipsoidal approximation becomes very straight and simple. In above analysis, we use the normalized frequency, and thus the passband becomes $\overline{\omega} \in [0, 1]$.

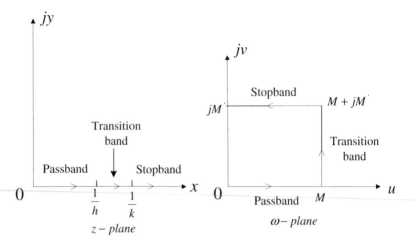

Fig. 7.3 Tri-band correspondence in conformal mapping of $z = s_l(\omega)$

Table 7.1 Construction comparison of the approximation functions

Chebyshev approximation	Construction form	Ellipsoidal approximation
$\cos[n \cos^{-1}(\overline{\omega})]$	function $[f(n)(\text{inverse function})]$	$S_l\left[\frac{nM_1}{M} s_l^{-1}(\overline{\omega}), k_1\right]$

Case 1 Transformation of passband
   Assuming

$$J = S_l^{-1}(\overline{\omega}, k, h) \tag{7.9}$$

$$l = \left(\frac{nM_1}{M}\right)J \tag{7.10}$$

We obtain

$$\begin{array}{ll} \overline{\omega} \text{ plane} & \text{variation} : -\frac{1}{h} \to \frac{1}{h} \\ J \text{ plane} & \text{variation} : -M \to M \\ l \text{ plane} & \text{variation} : -nM_1 \to nM_1 \end{array} \tag{7.11}$$

Note that

$$J = S_l^{-1}(\overline{\omega}, k, h) = \int_0^{\overline{\omega}} \frac{dt}{\sqrt{1 - h^2 t^2}\sqrt{1 - k^2 t^2}} \tag{7.12}$$

Figure 7.4 shows the ellipsoidal function transformation.

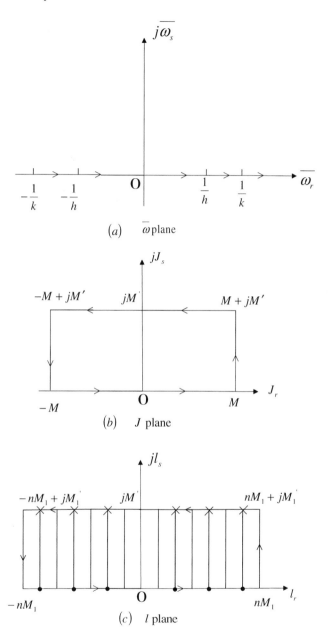

Fig. 7.4 Ellipsoidal function transformation

In Fig. 7.4 n zeros are represented by symbol •. Here without loss of generality, n is set as odd number. Hence, the locations of the zeros can be written as

$$\frac{nM_1}{M}J = 2mM_1, \quad m = 0, \pm 1, \cdots \pm \frac{1}{2}(n-1) \tag{7.13}$$

$$J = S_l^{-1}(\overline{\omega}, k, h) = \frac{2m}{n}M \tag{7.14}$$

Finally, the locations of m zeros can be uniformly expressed as

$$\overline{\omega}_{0m} = S_l\left(\frac{2m}{n}, k, h\right) \tag{7.15}$$

Case 2 Transformation of transition band

$\overline{\omega}$ plane	variation : $\frac{1}{h} \to \frac{1}{k}$
J plane	variation : $M \to M + jM'$
l plane	variation : $nM_1 \to nM_1 + j\left(\frac{nM_1M'}{M}\right)$

$$\tag{7.16}$$

The necessary requirement for the transition band is that l-plane correspondence must ensure the monotonous decrease of the transient band. Hence we have

$$\frac{nM_1M'}{M} = M_1' \tag{7.17}$$

A restriction condition on n can be obtained as

$$n = \frac{MM_1'}{M_1M'} \tag{7.18}$$

Case 3 Transformation of stopband

$\overline{\omega}$ plane	variation : $\frac{1}{k} \to \infty$
J plane	variation : $M \to M + jM'$
l plane	variation : $nM_1 \to nM + jM_1' \to jM_1'$

$$\tag{7.19}$$

According to the property of the ellipsoidal function, which is analogy to the elliptical function, there are $(n+1)/2$ poles in the region. Specifically,

$$\frac{nM_1}{M}J = 2mM_1 + jM_1' \tag{7.20}$$

Finally, the locations of the m poles can be uniformly as

$$\overline{\omega}_{\infty m} = S_l\left(\frac{2m}{n}M + jM', k, h\right) \tag{7.21}$$

In the following, we will give a lemma and a theorem.

Lemma

$$s_l(u + jM') = \frac{1}{khs_l(u)} \tag{7.22}$$

Proof we first, write a general addition formulation as

$$s_l(u + jv) = \frac{s_l(u)d_l(v) + jd_l(u)e_l(u)s_l(v)e_l(v)}{e_l^2(v) + k^2h^2s_l^2(u)s_l^2(v)} \tag{7.23}$$

Note that in writing the above formulation, we use

$$s_l(u) = s_l(u, k, h) \tag{7.24}$$

$$s_l(v) = s_l\left(v, \widetilde{k}, h\right) \tag{7.25}$$

where

$$\widetilde{k} = \sqrt{h^2 - k^2} \tag{7.26}$$

Now we make a concrete discussion on

$$\begin{aligned} s_l[u + (M + jM')] &= s_l(u + v) \\ &= \frac{s_l(u)d_l(M + jM')e_l(M + jM') + s_l(M + jM')d_l(u)e_l(u)}{1 - k^2h^2s_l^2(u)s_l^2(M + jM')} \end{aligned} \tag{7.27}$$

Note that according to conformal mapping of the ellipsoidal function as shown in Fig. 6.3, we can clearly see the following relationship

$$s_l(M + jM') = \frac{1}{k} \tag{7.28}$$

$$d_l(M + jM') = \sqrt{1 - k^2s_l^2(M + jM')} = 0 \tag{7.29}$$

$$e_l(M + jM') = \sqrt{1 - h^2s_l^2(M + jM')} = -j\sqrt{\frac{h^2 - k^2}{k^2}} = -j\frac{\widetilde{k}}{k} \tag{7.30}$$

Substituting Eqs. (7.28–7.30) into Eq. (7.23), we have

$$s_l[u + (M + jM')] = \frac{\frac{1}{k}d_l(u)e_l(u)}{1 - h^2s_l^2(u)} = \frac{d_l(u)e_l(u)}{ke_l^2(u)}$$

Finally, we can get

$$s_l[u + (M + jM')] = \frac{d_l(u)}{ke_l(u)} \tag{7.31}$$

On the other hand, we turn our attention on $s_l[u + M]$. Consider

$$s_l[u + M] = s_l[u + v] = \frac{s_l(u)d_l(M)e_l(M) + d_l(u)e_l(u)s_l(M)}{1 - k^2 h^2 s_l^2(u)s_l^2(M)} \tag{7.32}$$

According to conformal mapping again, we can know

$$s_l(M) = \frac{1}{h} \tag{7.33}$$

$$d_l(M) = \frac{\tilde{k}}{h} \tag{7.34}$$

$$e_l(M) = 0 \tag{7.35}$$

Inserting the above formulations into Eq. (7.32), we have

$$s_l[u + M] = \frac{\frac{1}{h}d_l(u)e_l(u)}{1 - k^2 s_l^2(u)} = \frac{e_l(u)}{hd_l(u)} \tag{7.36}$$

Similarly, we can obtain

$$d_l[u + M] = \frac{g'}{d_l(u)} \tag{7.37}$$

$$e_l[u + M] = -\frac{hg's_l(u)}{d_l(u)} \tag{7.38}$$

Now we rewrite Eq. (7.36) as

$$s_l[(u + M) + jM'] = \frac{s_l(u + M)d_l(jM')e_l(jM') + d_l(u + M)e_l(u + M)s_l(jM')}{1 - k^2 h^2 s_l^2(u + M)s_l^2(jM')} \tag{7.39}$$

From Fig. 6.3, we can easily obtain

$$s_l(jM') \rightarrow \infty \tag{7.40}$$

$$d_l(jM') \rightarrow jks_l(jM') \tag{7.41}$$

$$e_l(jM') \rightarrow jhs_l(jM') \tag{7.42}$$

Substituting the above conditions into Eq. (7.39), we have

$$s_l[(u+M)+jM'] = \frac{-khs_l(u+M)}{-k^2h^2s_l^2(u+M)}$$

Hence we can get

$$s_l[(u+M)+jM'] = \frac{1}{khs_l(u+M)} \tag{7.43}$$

Theorem *the relationship between zeros and poles in the ellipsoidal-function response is*

$$\overline{\omega}_{\infty m} = \frac{1}{khs_l\left(\frac{2m}{n}M, k, h\right)} = \frac{1}{kh\overline{\omega}_{0m}} \tag{7.44}$$

Based on the above Lemma, the theorem can be easily proved.

7.3   Ellipsoidal Function Filter

Here, we give a detailed discussion from three perspectives as follows.

7.3.1   *Frequency Response Function of the Filter*

We use the zeros of the ellipsoidal function $\overline{\omega}_{0m}$ and $\overline{\omega}_{\infty m}$ to give the approximation polynomial of the filter response as

$$F_n(\overline{\omega}) = A_0 \frac{\overline{\omega}\left(\overline{\omega}_1^2 - \overline{\omega}^2\right)\left(\overline{\omega}_2^2 - \overline{\omega}^2\right)\cdots\left(\overline{\omega}_p^2 - \overline{\omega}^2\right)}{\left(1 - k^2h^2\overline{\omega}_1^2\overline{\omega}^2\right)\left(1 - k^2h^2\overline{\omega}_2^2\overline{\omega}^2\right)\cdots\left(1 - k^2h^2\overline{\omega}_p^2\overline{\omega}^2\right)} \tag{7.45}$$

Note that in Eq. (7.45) n is considered as an odd number without loss of generality. The corresponding zeros are

$$\overline{\omega}_m = s_l\left(\frac{2m}{n}M\right) \tag{7.46}$$

in which

$$m = 1, 2, \cdots p = \frac{1}{2}(n - 1) \tag{7.47}$$

Case 1 Edge condition of the filter

At the edge of operational frequency band of the filter, we have

$$\overline{\omega} = \frac{1}{h}, \quad F_n(1) = s_l(nM') = \frac{1}{h}(-1)^p \tag{7.48}$$

Substituting Eq. (7.48) into Eq. (7.45), we obtain

$$\frac{(-1)^p}{h} = A_0 \frac{\frac{1}{h}\left(\overline{\omega}_1^2 - \frac{1}{h^2}\right)\left(\overline{\omega}_2^2 - \frac{1}{h^2}\right)\cdots\left(\overline{\omega}_p^2 - \frac{1}{h^2}\right)}{\left(1 - k^2\overline{\omega}_1^2\right)\left(1 - k^2\overline{\omega}_2^2\right)\cdots\left(1 - k^2\overline{\omega}_p^2\right)} \tag{7.49}$$

Case 2 Transition condition of the filter

For the transition band, we have

$$\overline{\omega} = \frac{1}{k}, \quad F_n\left(\frac{1}{k}\right) = \frac{1}{k_1}(-1)^p \tag{7.50}$$

In Eq. (7.50) the following equation has been used, i.e.,

$$s_l\left(nM_1 + jM_1', k_1\right) = \frac{(-1)^p}{k_1} \tag{7.51}$$

Further, we can write

$$\frac{(-1)^p}{k_1} = A_0 \frac{\frac{1}{k}\left(\overline{\omega}_1^2 - \frac{1}{k^2}\right)\left(\overline{\omega}_2^2 - \frac{1}{k^2}\right)\cdots\left(\overline{\omega}_p^2 - \frac{1}{k^2}\right)}{\left(1 - h^2\overline{\omega}_1^2\right)\left(1 - h^2\overline{\omega}_2^2\right)\cdots\left(1 - h^2\overline{\omega}_p^2\right)}$$
$$= \frac{A_0}{k^n} \frac{\left(k^2\overline{\omega}_1^2 - 1\right)\left(k^2\overline{\omega}_2^2 - 1\right)\cdots\left(k^2\overline{\omega}_p^2 - 1\right)}{\left(1 - h^2\overline{\omega}_1^2\right)\left(1 - h^2\overline{\omega}_2^2\right)\cdots\left(1 - h^2\overline{\omega}_p^2\right)} \tag{7.52}$$

Equation (7.49) can be reexpressed as

$$\frac{(-1)^p}{h} = \frac{A_0}{h^n} \frac{\left(h^2\overline{\omega}_1^2 - 1\right)\left(h^2\overline{\omega}_2^2 - 1\right)\cdots\left(h^2\overline{\omega}_p^2 - 1\right)}{\left(1 - k^2\overline{\omega}_1^2\right)\left(1 - k^2\overline{\omega}_2^2\right)\cdots\left(1 - k^2\overline{\omega}_p^2\right)} \tag{7.53}$$

By multiplying Eq. (7.52) by Eq. (7.53), we can get

$$A_0 = \left(\frac{k^n h^{n-1}}{k_1}\right)^{\frac{1}{2}}$$

(7.54)

By dividing Eq. (7.52) by Eq. (7.53), we have

$$\frac{h}{k_1} = \frac{h^n}{k^n}\left[\frac{\left(1-k^2\overline{\omega}_1^2\right)\left(1-k^2\overline{\omega}_2^2\right)\cdots\left(1-k^2\overline{\omega}_p^2\right)}{\left(1-h^2\overline{\omega}_1^2\right)\left(1-h^2\overline{\omega}_2^2\right)\cdots\left(1-h^2\overline{\omega}_p^2\right)}\right]^2$$

(7.55)

Equation (7.55) can be rewritten as

$$\sqrt{\frac{1}{k_1}} = \sqrt{\frac{h^{n-1}}{k^n}}\left[\frac{\left(1-k^2\overline{\omega}_1^2\right)\left(1-k^2\overline{\omega}_2^2\right)\cdots\left(1-k^2\overline{\omega}_p^2\right)}{\left(1-h^2\overline{\omega}_1^2\right)\left(1-h^2\overline{\omega}_2^2\right)\cdots\left(1-h^2\overline{\omega}_p^2\right)}\right]$$

(7.56)

Substituting Eq. (7.56) into Eq. (7.54), the response of the filter $F_n(\overline{\omega})$ can e-expressed as

$$F_n(\overline{\omega}) = h^{n-1}\frac{\overline{\omega}\left(\overline{\omega}_1^2-\overline{\omega}^2\right)\left(\overline{\omega}_2^2-\overline{\omega}^2\right)\cdots\left(\overline{\omega}_p^2-\overline{\omega}^2\right)}{\left(1-k^2h^2\overline{\omega}_1^2\overline{\omega}^2\right)\left(1-k^2h^2\overline{\omega}_2^2\overline{\omega}^2\right)\cdots\left(1-k^2h^2\overline{\omega}_p^2\overline{\omega}^2\right)}$$

$$\cdot\,\frac{\left(1-k^2\overline{\omega}_1^2\right)\left(1-k^2\overline{\omega}_2^2\right)\cdots\left(1-k^2\overline{\omega}_p^2\right)}{\left(1-h^2\overline{\omega}_1^2\right)\left(1-h^2\overline{\omega}_2^2\right)\cdots\left(1-h^2\overline{\omega}_p^2\right)}$$

(7.57)

7.3.2   Synthesis of the Ellipsoidal Function Filter

Figure 7.5 shows the response of the ellipsoidal function.

Case 1 Passband response $\overline{\omega} \in \left[0, \frac{1}{h}\right]$

The gain $G\left(\overline{\omega}^2\right)$ can be expressed as

$$G\left(\overline{\omega}^2\right) = \frac{H_n}{1+\varepsilon^2 F_n^2(\omega)}$$

(7.58)

From Fig. 7.5 it can be seen that

$$G_{\max} = H_n$$

(7.59)

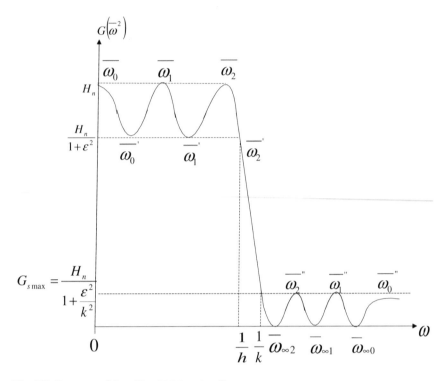

Fig. 7.5 Response of the ellipsoidal function filter

$$G_{\min} = \frac{H_n}{1 + \varepsilon^2} \qquad (7.60)$$

In the passband, the filter has an equal-ripple characteristic. $G_{\max}$ is obtained at $\overline{\omega}_m$, e.g.,

$$\begin{cases} \overline{\omega}_m = s_l\left(\frac{2m}{n}M\right) \\ m = 0, 1, 2, \cdots \frac{1}{2}(n-1) \\ n \text{ is odd number} \end{cases} \qquad (7.61)$$

and also $G_{\min}$ is gotten at $\overline{\omega}'_m$

$$\begin{cases} \overline{\omega}'_m = s_l\left(\frac{2m-1}{n}M\right) \\ m = 1, 2, \cdots \frac{1}{2}(n-1) \end{cases} \qquad (7.62)$$

Case 2 Transition band response $\overline{\omega} \in \left[\frac{1}{h}, \frac{1}{k}\right]$

Gain of the filter $G(\overline{\omega}^2)$ decreases from the maximum $G_{\min} = H_n/(1+\varepsilon^2)$ to

$$G_{s\max} = \frac{H_n}{1 + \frac{\varepsilon^2}{k_1^2}} \tag{7.63}$$

Here the descent depends on slope k.

Case 3 Stopband response $\overline{\omega} \in [\frac{1}{k}, \infty)$

In this case $G(\overline{\omega}^2) = 0$, which corresponds to poles of the gain

$$\overline{\omega}_{\infty m} = \frac{1}{hk\overline{\omega}_m} \tag{7.64}$$

On the other hand, the frequency related to the maximum of the gain $G(\overline{\omega}^2) = G_{s\max}$ is

$$\overline{\omega}_m'' = \frac{1}{hk\overline{\omega}_m'} \tag{7.65}$$

Synthesis procedure of the ellipsoidal function filter is shown in Figs. 7.6 and 7.7.

Three important performance parameters
Passband ripple — $\varepsilon = \sqrt{10^{\frac{AP}{10}} - 1}$, $AP = 10\log(1 + \varepsilon^2)\, dB$ Frequency — passband $\frac{1}{h}$ and edge of the band $\frac{1}{k}$ Minimal complex region out of the band — $AS = 10\log\left(1 + \frac{\varepsilon^2}{k_1^2}\right) - AP \ \ dB$

Five filter parameters to be synthesized
n —related to the number of the components, but unnecessary to be equal to the number of the components; h —related to the passband; ε —related to the ripple in the passband; k —related to the slope in the transition band; k_1 —related to the maximal gain in the stopband $G_{s\max}$;

Fig. 7.6 Ellipsoidal function synthesis from the performance to the parameters of the filter

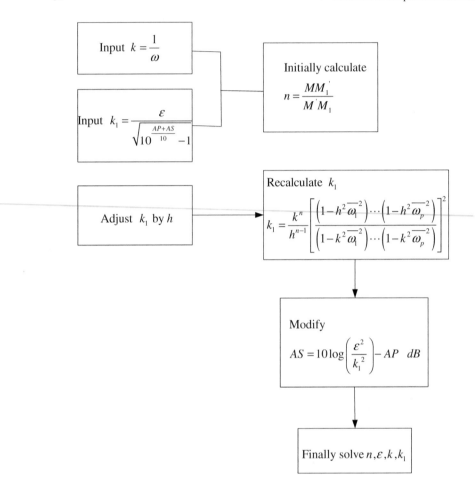

Fig. 7.7 Solution of the filter parameters $\overline{\omega}, \varepsilon, AS$ and AP

7.3.3 Ellipsoidal Function Filter with n = 3

We take the ellipsoidal function with $n = 3$ as an example for the ellipsoidal function filter. Figure 7.8 shows the filter response.

Introducing the complex frequency plane S, we have

$$\overline{s} = j\overline{\omega} \tag{7.66}$$

We investigate the filter synthesis of the ellipsoidal function with $n = 3$. In this case, $p = (n-1)/2 = 1$, and we can obtain

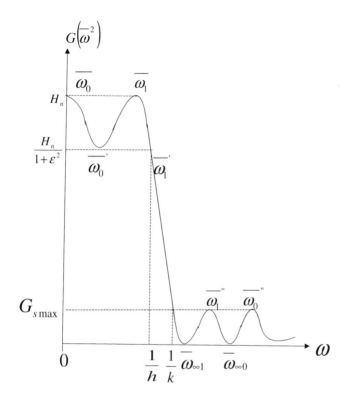

Fig. 7.8 Ellipsoidal function filter response with $n = 3$

$$F_3^2(-js) = -A_0^2 \frac{\bar{s}^2(\bar{s}^2 + \overline{\omega}_1^2)^2}{(1 + k^2 h^2 \bar{s}^2 \overline{\omega}_1^2)^2} \qquad (7.67)$$

With the use of the scattering network parameters S and its energy conservation property for the lossless network, we have

$$S_{11}(\bar{s})S_{11}(-\bar{s}) = 1 - S_{21}(\bar{s})S_{21}(-\bar{s})$$
$$= 1 - G(-\bar{s}^2) = \frac{\varepsilon^2 F_3^2(-j\bar{s})}{1 + \varepsilon^2 F_3^2(-j\bar{s})} \qquad (7.68)$$

We can easily write

$$S_{11}(\bar{s})S_{11}(-\bar{s}) = \frac{\bar{s}^2(\bar{s}^2 + \overline{\omega}_1^2)}{(\bar{s} - \bar{s}_0)(\bar{s} + \bar{s}_0)(\bar{s} - \bar{s}_1)(\bar{s} - \bar{s}_{-1})(\bar{s} - \bar{s}_2)(\bar{s} - \bar{s}_{-2})} \qquad (7.69)$$

in which $\pm \bar{s}_0, \bar{s}_1, \bar{s}_{-1}, \bar{s}_2$ and $\bar{s}_{-2}$ are the roots of the equation

$$1 + \varepsilon^2 F_3^2(-j\bar{s}) = 0 \tag{7.70}$$

Specifically,

$$F_3(-j\bar{s}) = \pm j\frac{1}{\varepsilon} \tag{7.71}$$

Thus we have

$$s_l\left[\frac{nM_1}{M} s_l^{-1}(-j\bar{s}, k), k_1\right] = \pm j\frac{1}{\varepsilon} \tag{7.72}$$

One of the roots is

$$s_l(jv, k_1) = j\frac{1}{\varepsilon} \tag{7.73}$$

According to (7.73), we can have

$$s_l(jv, k_1) = j\frac{s_l(v, \tilde{k}, h)}{e_l(v, \tilde{k}, h)} = j\frac{s_l(v, \tilde{k}, h)}{\sqrt{1 - h^2 s_l^2(v, \tilde{k}, h)}} = j\frac{1}{\varepsilon} \tag{7.74}$$

Here $\tilde{k} = \sqrt{h^2 - k^2}$. Further we can easily get

$$s_l(v, \tilde{k}, h) = \frac{1}{\sqrt{h^2 + \varepsilon^2}} \tag{7.75}$$

We calculate v according to the definition of the ellipsoidal integral, e.g.,

$$\frac{nM_1}{M} s_l^{-1}(-j\bar{s}, k) = \pm jv + 2mM_1 \tag{7.76}$$

$$m = 0, \pm 1 \tag{7.77}$$

From (7.76), the roots can be solved as

$$\bar{s}_m = js_l\left(\frac{2m}{n}M + j\frac{M}{nM_1}v, k\right) \tag{7.78}$$

Specifically, we have

$$\bar{s}_0 = js_l\left(j\frac{M}{3M_1}v, k\right) \tag{7.79}$$

$$\bar{s}_{\pm 1} = j s_l \left(j \frac{M}{3M_1} v \pm \frac{2}{3} M, k \right) \tag{7.80}$$

Assigning the roots in the half left plane $\bar{s}$ to $s_{11}(\bar{s})$, we have

$$S_{11}(\bar{s}) = \frac{\bar{s}(\bar{s}^2 + \bar{\omega}_1^2)}{(\bar{s} - \bar{s}_0)(\bar{s} - \bar{s}_1)(\bar{s} - \bar{s}_{-1})} \tag{7.81}$$

On the other hand, the network of the ellipsoidal function filter is illustrated in Fig. 7.9.

Hence, the normalized input impedance $\bar{Z}_{in}(\bar{s})$ can be calculated as

$$\bar{Z}_{in}(\bar{s}) = \bar{L}_1 \bar{S} + \frac{1}{\frac{1}{1 + \bar{L}_1 \bar{S}} + \frac{1}{\bar{L}_2 \bar{S} + \frac{1}{\bar{C}_2 \bar{S}}}} \tag{7.82}$$

Equation (7.82) can be reduced to

$$\bar{Z}_{in}(\bar{s}) = \frac{\bar{L}_1(\bar{L}_1 + 2\bar{L}_2)\bar{C}_2 \bar{S}^3 + (\bar{L}_1 + \bar{L}_2)\bar{C}_2 \bar{S}^2 + 2\bar{L}_1 \bar{S} + 1}{(\bar{L}_1 + \bar{L}_2)\bar{C}_2 \bar{S}^2 + \bar{C}_2 \bar{S} + 1} \tag{7.83}$$

According to relation between the normalized input impedance $\bar{Z}_{in}(\bar{s})$ and the scattering parameters S

$$\bar{Z}_{in}(\bar{s}) = \frac{1 + S_{11}(\bar{s})}{1 - S_{11}(\bar{s})} \tag{7.84}$$

and assuming

$$S_{\pm 1} = SR \pm jSI \tag{7.85}$$

$$\omega_1 = \bar{\omega}_1 \tag{7.86}$$

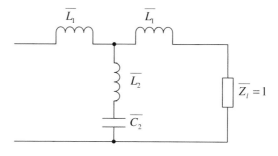

Fig. 7.9 Network of the ellipsoidal function filter with $n = 3$

we have

$$\overline{Z}_{in}(\overline{s}) = \frac{\frac{2}{\overline{S}_0(SR^2+SI^2)}\overline{S}^3 + \frac{\overline{S}_0+2SR}{\overline{S}_0(SR^2+SI^2)}\overline{S}^2 + \frac{2\overline{S}_0SR+SR^2+SI^2+\omega_1^2}{\overline{S}_0(SR^2+SI^2)}\overline{S}+1}{\frac{\overline{S}_0+2SR}{\overline{S}_0(SR^2+SI^2)}\overline{S}^2 + \frac{2\overline{S}_0SR+SR^2+SI^2-\omega_1^2}{\overline{S}_0(SR^2+SI^2)}\overline{S}+1}$$

$$= \frac{2\overline{S}^3 + (\overline{S}_0+2SR)\overline{S}^2 + (2\overline{S}_0SR+SR^2+SI^2+\omega_1^2)\overline{S}+\overline{S}_0(SR^2+SI^2)}{(\overline{S}_0+2SR)\overline{S}^2 + (2\overline{S}_0SR+SR^2+SI^2-\omega_1^2)\overline{S}+\overline{S}_0(SR^2+SI^2)}$$

$$(7.87)$$

Finally, the parameters of the components in the network can be obtained

$$\overline{L}_1 = \frac{2\overline{S}_0SR+SR^2+SI^2+\omega_1^2}{\overline{S}_0(SR^2+SI^2)} \qquad (7.88)$$

$$\overline{C}_2 = \frac{2\overline{S}_0SR+SR^2+SI^2-\omega_1^2}{\overline{S}_0(SR^2+SI^2)} \qquad (7.89)$$

$$\overline{L}_2 = \frac{\overline{S}_0+2SR}{\overline{S}_0(SR^2+SI^2)\overline{C}_2} - \overline{L}_1 \qquad (7.90)$$

As shown in Fig. 7.9, when $n = 3$, the number of the components of the ellip-soidal function filter is 4.

Appendix A
Correlation of Two Kinds of Ellipsoidal Integrals

Complete ellipsoidal integral of the first kind and general elliptic integral of the first kind satisfy

$$G(k,h) = \frac{1}{h}u\left(\frac{k}{h},h\right)$$

This can strongly manifest that: the relationship is extremely close between complete ellipsoidal integral of the first kind and general elliptic integral of the first kind. The definition of $G(k,h)$ is known as

$$G(k,h) = \int_0^1 \frac{dx}{\sqrt{1-h^2x^2}\sqrt{1-k^2x^2}} \tag{A.1}$$

where

$$\begin{cases} k = \frac{\sqrt{a^2-b^2}}{a} \\ h = \frac{\sqrt{a^2-c^2}}{a} \end{cases} \tag{A.2}$$

Without loss of generality, considering $a \geq b \geq c$, we have

$$k \leq h \tag{A.3}$$

Now assuming

$$y = hx \tag{A.4}$$

© Science Press, Beijing and Springer Nature Singapore Pte Ltd. 2017
C.-H. Liang, *Notes on the Ellipsoidal Function*,
DOI 10.1007/978-981-10-2908-0

we can get

$$
\begin{cases}
x = \frac{y}{h} \\[4pt]
dx = \frac{1}{h} dy \\[4pt]
x \in [0, 1] \sim y \in [0, h]
\end{cases}
\tag{A.5}
$$

Equation (A.1) can be rewritten as

$$
G(k, h) = \frac{1}{h} \int_0^h \frac{dy}{\sqrt{1 - y^2}\sqrt{1 - \left(\frac{k}{h}\right)^2 y^2}}
\tag{A.6}
$$

Namely

$$
G(k, h) = \frac{1}{h} u\left(\frac{k}{h}, h\right)
\tag{A.7}
$$

Note that

$$
x = \frac{k}{h} \leq 1
\tag{A.8}
$$

which meets the conditions for the definition of the elliptic integrals.

Some people may ask, "Why should we devise the ellipsoidal integrals in such a case?" There are two main reasons:

1. Elliptic integrals deal with essentially two-dimensional problems, with only one independent parameter, namely k; and ellipsoidal integrals can deal with essentially three-dimensional problems, with two independent parameters, namely k and h.
2. $G(k, h)$ is a complete ellipsoidal integral of the first kind and can be expanded into a general ellipsoidal integral of the first kind, namely

$$
m(x, k, h).
\tag{A.9}
$$

Appendix B
Correlation of $E(\xi)$ and $n(x,e,h)$

The complete elliptic integral of the second kind $E(\xi)$ and general ellipsoidal integral of the second kind $n(x,e,h)$ meet

$$\frac{\mathrm{d}n(x,e,h)}{\mathrm{d}x} = \sqrt{1-h^2x^2}\,E\left(\frac{e\sqrt{1-x^2}}{\sqrt{1-h^2x^2}}\right)$$

The relationship is very close between complete elliptic integral of the second kind $E(\xi)$ and general ellipsoidal integral of the second kind $n(x,e,h)$.

It is known that general ellipsoidal integral of the second kind is defined as

$$n(x,e,h) = \int_0^x \sqrt{1-h^2t^2}\,E\left(\frac{e\sqrt{1-t^2}}{\sqrt{1-h^2t^2}}\right)\mathrm{d}t \qquad (B.1)$$

where

$$\begin{cases} e = \frac{\sqrt{b^2-c^2}}{b} \\ h = \frac{\sqrt{a^2-c^2}}{a} \end{cases} \qquad (B.2)$$

The right side of Eq. (B.1) can be considered as an integral with parameter x. We take the derivative of Eq. (B.1) with respect to x as

$$\frac{\mathrm{d}n(x,e,h)}{\mathrm{d}x} = \sqrt{1-h^2x^2}\,E\left(\frac{e\sqrt{1-x^2}}{\sqrt{1-h^2x^2}}\right)$$

$$= \sqrt{1-h^2x^2}\,E(\xi) \qquad (B.3)$$

© Science Press, Beijing and Springer Nature Singapore Pte Ltd. 2017 91
Z.-H. Liang, *Notes on the Ellipsoidal Function*,
DOI 10.1007/978-981-10-2908-0

where

$$\xi = \frac{e\sqrt{1-x^2}}{\sqrt{1-h^2x^2}}$$
(B.4)

This happens to be the direct connection between the complete elliptic integral of the second kind $E(\xi)$ and the general ellipsoidal integral of the second kind $n(x,e,h)$.

Appendix C
Other form of M

The integral M can be expressed as

$$M = \int_0^{\frac{1}{h}} \frac{dt}{\sqrt{1 - h^2 t^2}\sqrt{1 - k^2 t^2}}$$
$$= G(k, h) + F'(p', h')$$

Here we write M in terms of complete ellipsoidal integrals adding an extra term $F'(p', h')$. Note that

$$F'(p', h') = \int_1^{\frac{1}{h}} \frac{dt}{\sqrt{1 - h^2 t^2}\sqrt{1 - k^2 t^2}} \tag{C.1}$$

To realize the transformation

$$\begin{cases} t = 1 \leftrightarrow t_1 = 0 \\ t = \frac{1}{h} \leftrightarrow t_1 = 1 \end{cases} \tag{C.2}$$

we use the following relationship

$$t = \frac{1}{\sqrt{1 - h'^2 t_1^2}} \tag{C.3}$$

in which

$$h' = \sqrt{1 - h^2} \tag{C.4}$$

© Science Press, Beijing and Springer Nature Singapore Pte Ltd. 2017
C.-H. Liang, *Notes on the Ellipsoidal Function*,
DOI 10.1007/978-981-10-2908-0

And we have

$$dt = \frac{h'^2 t_1 dt_1}{(1 - h'^2 t_1^2)^{3/2}} \tag{C.5}$$

$$1 - h^2 t^2 = 1 - \frac{h^2}{1 - h'^2 t_1^2}$$
$$= \frac{1 - h'^2 t_1^2 - h^2}{1 - h'^2 t_1^2} = \frac{h'^2 (1 - t_1^2)}{(1 - h'^2 t_1^2)}$$

Hence we can obtain

$$\sqrt{1 - h^2 t^2} = \frac{h' \sqrt{1 - t_1^2}}{\sqrt{1 - h'^2 t_1^2}} \tag{C.6}$$

and

$$1 - k^2 t^2 = 1 - \frac{k^2}{1 - h'^2 t_1^2}$$
$$= \frac{1 - k^2 - h'^2 t_1^2}{1 - h'^2 t_1^2} = \frac{k'^2 - h'^2 t_1^2}{1 - h'^2 t_1^2}$$
$$= \frac{k'^2 [1 - \left(\frac{h'}{k'}\right)^2 t_1^2]}{1 - h'^2 t_1^2} \tag{C.7}$$

Introducing a new variable

$$p' = \left(\frac{h'}{k'}\right) \tag{C.8}$$

and noticing

$$\left(\frac{h'}{k'}\right) \neq \left(\frac{h}{k}\right)' \tag{C.9}$$

$$\left(\frac{h'}{k'}\right) \leq 1 \tag{C.10}$$

we have

$$\sqrt{1 - k^2 t^2} = \frac{k' \sqrt{1 - p'^2 t_1^2}}{\sqrt{1 - h'^2 t_1^2}} \tag{C.11}$$

Substituting (C.5), (C.6) and (C.11) into the expression of the integral M, we get

$$
\int_{1}^{\frac{1}{h}} \frac{dt}{(1 - h^2 t^2)^{1/2}(1 - k^2 t^2)^{1/2}} = \int_{0}^{1} \frac{\frac{h'^2 t_1 dt_1}{(1 - h'^2 t^2)^{3/2}}}{\frac{h'\sqrt{1 - t_1^2}}{(1 - h'^2 t_1^2)^{1/2}} \cdot \frac{k'\sqrt{1 - p'^2 t_1^2}}{(1 - h'^2 t_1^2)^{1/2}}}
$$

$$
= \int_{0}^{1} \frac{p' t_1 dt_1}{\sqrt{1 - t_1^2}\sqrt{1 - h'^2 t_1^2}\sqrt{1 - p'^2 t_1^2}} \qquad \text{(C.12)}
$$

With a new integral

$$
F'(p', h') = \int_{0}^{1} \frac{p' t_1 dt_1}{\sqrt{1 - t_1^2}\sqrt{1 - h'^2 t_1^2}\sqrt{1 - p'^2 t_1^2}} \qquad \text{(C.13)}
$$

finally we can obtain

$$
M = G(k, h) + F'(p', h') \qquad \text{(C.14)}
$$

In (C.14), the integral M is written as the summation of the complete ellipsoidal integrals adding an extra term.

References

Chang-Hong Liang, *Computational Microwave*, Xi'an: Northwest Institute of Telecommunication Engineering Press, 1985.

Chang-Hong Liang, *Notes on Complex Variable Function*, Beijing: Science Press, 2012.

Zhu-Xi Wang and Dun-Ren Guo, *An Introduction to Special Function*, Beijing: Science Press, 1979.

Qi-Xiao Ye and Yong-Huan Shen, *Practical Mathematical Handbook*, Beijing: Science Press, 1992.

© Science Press, Beijing and Springer Nature Singapore Pte Ltd. 2017
C.-H. Liang, *Notes on the Ellipsoidal Function*,
DOI 10.1007/978-981-10-2908-0

Printed in the United States
By Bookmasters